TELEMEDICINE

MEDICINE AND COMMUNICATION

TELEMEDICINE
MEDICINE AND COMMUNICATION

Edited by

Thorsten M. Buzug
University of Applied Sciences Kolbenz
RheinAhrCampus Remagen, Germany

Heinz Handels
Medical University Lübeck
Lübeck, Germany

and

Dietrich Holz
University of Applied Sciences Kolbenz
RheinAhrCampus Remagen, Germany

Springer Science+Business Media, LLC

Library of Congress Cataloging-in-Publication Data

Telemedicine: medicine and communication/edited by Thorsten M. Buzug, Heinz Handels, and Dietrich Holz.
p. cm.
"Proceedings of the International Conference on Emerging Technologies and Life Science: Medicine and Communication (MEDICOM 2000), held September 28–29, 2000, in Remagen, Germany."
Includes bibliographical references and index.
ISBN 978-0-306-46552-9 ISBN 978-1-4615-1253-0 (eBook)
DOI 10.1007/978-1-4615-1253-0

1. Telecommunication in medicine—Congresses. 2. Medical telematics—Congresses. I. Buzug, Thorsten M., 1963- II. Handels, Heinz. III. Holz, Dietrich, 1956- IV. International Conference on Emerging Technologies and Life Sciences: Medicine and Communication (MEDICOM 2000) (2000: Remagen, Germany)

R119.9 .T4515 2001

2001020265

Proceedings of the International Conference on Emerging Technologies and Life Sciences: Medicine and Communciation (MEDICOM 2000), held September 28–29, 2000, in Remagen, Germany. The Conference was part of the Bonn Science Festival organized by the Strukturförderungsgesellschaft mbH (SFG).

ISBN 978-0-306-46552-9

Originally published by Kluwer Academic / Plenum Publishers, New York in 2001

http://www.wkap.nl/

10 9 8 7 6 5 4 3 2 1

A C.I.P. record for this book is available from the Library of Congress.

Preface

Within the various aspects of life-science technologies medicine and information technology will change next millennium's quality-of-life fundamentally. Thanks to the rapid growth of telecommunication industry and the success and popularity of the internet the face of medicine will essentially change, because information technology is expected to play a major role in future health care systems.

The conference MEDICOM 2000 is a discussion forum on fast and cost efficient patient-data exchange systems between doctors' offices, medical laboratories, telearchive services, health care insurances, highly specialized experts in hospitals etc. The conference brought together scientific, medical and application experts from university, clinical and commercial sites of both areas - medicine and communication - to stimulate synergy between these rapidly evolving future technologies.

We would like to acknowledge all the parties who contributed to the success of the conference. Especially, we would like to thank Gisela Niedzwetzki and Waltraud Ott for secretarial support as well as Dirk Thomsen for web mastering. Additionally, we have to acknowledge the valuable support of Holger Dörle, Thomas Giese, Peter Just, Stefan Klöckner, Heike Lahr and Kerstin Lüdtke-Buzug during the conference.

We would also like to thank all the reviewers from the MEDICOM program committee for their valuable help in the reviewing process. Last but not least, we would like to thank all our exhibitors and sponsors - especially Medical Net, New Office and the Kreissparkasse Ahrweiler - who made the conference possible. We would like to thank the "Strukturförderungsgesellschaft Bonn/Rhein-Sieg/Ahrweiler mbH" - especially Gisela Ditze and Rolf Beyer - for accepting MEDICOM 2000 as part of the Sciences Festival Bonn.

Boards

ORGANIZATION COMMITTEE

Prof. Dr. **Thorsten M. Buzug** (Conference Chair)
Department of Mathematics and Technology
University of Applied Sciences Koblenz, RheinAhrCampus Remagen
buzug@rheinahrcampus.de

Prof. Dr. **Dictrich Holz** (Confercncc Co-Chair)
Department of Mathematics and Technology
University of Applied Sciences Koblenz, RheinAhrCampus Remagen
holz@rheinahrcampus.de

PD Dr. **Heinz Handels** (Chair Program Committee)
Institute of Medical Informatics
Medical University Lübeck
handels@medinf.mu-luebeck.de

Prof. Dr. **Andreas J. W. Goldschmidt**
Institute of Medical Biometrics, Informatics and Epidemiology
University of Bonn, Medical Center
andreas.goldschmidt@meb.uni-bonn.de

PROGRAM COMMITTEE

Opening Address

I would like to take this opportunity to thank you for inviting me to today's technical conference during Medicom 2000 here on the RheinAhrCampus in Remagen. I am extremely proud of this large scientific event here in the Ahrweiler district. The fact that Medicom 2000 is taking place here at "our Fachhochschule" underlines the growing significance of Ahrweiler as an important health and fitness region.

RheinAhrCampus is taking great strides in establishing itself in the region - as well as in the entire scientific community. The organizers have certainly backed the right horse with telemedicine by choosing a subject that has a bright future.

The world economy has witnessed dramatic change over the past few years. Market globalization and internationalization are occurring at breathtaking speed. This is just as true for the success of modern communication technologies. These are the driving forces behind economic growth, productivity and innovation. Innovative technologies are the main engine of growth and employment in Germany as well as throughout the world. Maintaining the lead through innovation - that is today's motto. The big fish no longer swallow up the little fish - instead the quick fish eat the slow.

This rapid development holds great promise for the health services, which are increasingly squeezed by enormous financial pressure. Information and telecommunication technologies offer many new and encouraging possibilities here; for example, I envision paperless doctor's offices and clinics, virtual networks of medical practices, or new forms of quality assurance.

Experts view telemedicine as the medicine of the next millennium - the new magic formula of modern healing. I feel that what many mockers refer to as "Nintendo surgery" offers great potential for the future. A patient can be in any operating room, for example here in Remagen, while a specialist on the other side of the world operates the scalpel via computer in San Francisco.

I challenge you to take advantage of the new mindset created by the new media and especially the Internet. Let us create a platform for telemedicine applications in the health services. Whether it is patient data, general medical knowledge or administrative files, we need an electronic health services network that meets national and international standards, embedded of course in a modern data security system.

There is a saying in English that whoever doesn't have time to take care of his health now will have to have time for his illness later. Since health is a person's most important asset, the health sector - apart from the information and telecommunication technologies - is becoming the boom industry of the future. What could be more natural than to combine these developments? In the district of Ahrweiler we have recognized this trend in good time and have set the course for the future. We are already way ahead of the race while others are still saddling the horses.

The RheinAhrCampus in Remagen is the flagship of Ahrweiler's health competence. The Fachhochschule is building on the strengths of our region with its innovative curricula. We can expect valuable synergy effects from our "Physical technologies" course of study with a focus on medical technology.

We are anticipating the beginning of construction of the Innovation and Start-up Center Sinzig (Innovations- und Gründerzentrum Sinzig or IGZ) in our immediate vicinity. The focus of the IGZ is also on health services, and one of its purposes is to ease the transitional phase for young entrepreneurs, who are fresh out of college and want to start their own businesses based on a future-oriented idea.

City planners are also intending to build a technology park in the county. They plan to attract companies from the future-oriented industries of biotechnology and health services, or more precisely medical technology, to the technology park.

The European Academy for the Study of Consequences of Scientific and Technological Advances in Bad Neuenahr-Ahrweiler is also contributing to the district of Ahrweiler's new profile.

The establishment of the Central Institute for Pharmaceutical Research and the Scientific Institute of the Pharmaceutical Manufacturers Research Association this year in Sinzig has made the "health web" in the district of Ahrweiler even more close-meshed.

These scientific facilities will be framed and linked together under the "Ahrweiler Health and Fitness Region" concept.

On the other hand, the educational and entrepreneurial opportunities in the district of Ahrweiler are a pillar in the overall structure of the Bonn/Rhein-Sieg/Ahrweiler region with all of its scientific facilities. We complement one another. We are all acting in concert. Together we are creating our own creative and attractive offer. We in the Bonn/Rhein-Sieg/Ahrweiler region have understood that the Europe of the future is a Europe of regions. Bonn/Rhein-Sieg/Ahrweiler is well on its way to becoming a region of science and research.

Through its economic and scientific endeavours, the district of Ahrweiler holds an excellent position in this competition, and today's conference at our Fachhochschule is the best proof of this. Ladies and gentlemen, let us continue on this path.

Landrat Dr. Jürgen Pföhler
September 28, 2000

Contents

TELE-IMAGING

Computer Supported Cooperative 3D Teleimaging in Java

HEINZ HANDELS, TOBIAS SCHÖßLER, JAN EHRHARDT, SIEGFRIED J. PÖPPL
Institute for Medical Informatics, Medical University of Lübeck, Ratzeburger Allee160, 23538 Lübeck, Germany.

Key words: CSCW, Java, 3D visualisation, teleimaging

Abstract: The software system Cypris enables the shared 2D and 3D visualisation of medical image sequences. During a 3D teleimaging session threedimensional visualisations of medical objects (organs, tumors, bones etc.) are displayed simultaneously on the screens of the teleconference partners. For the synchronisation of the telesessions the CSCW tool Jermes developed at our Institute is used. Jermes is a Java program library, in which functions for computer supported cooperative work are implemented. In a cooperative teleconference 3D objects can be manipulated and the 3D view can be changed interactively by both users. Due to the implemented communication concept low communication rates occur during a teleconference. Hence, Cypris can even be used via narrow-band networks like ISDN.

1. INTRODUCTION

The increasing availability of internet and intranet based communication structures created a variety of new possibilities for telemedical applications. Due to the high information content of medical image data exchange and telediscussion of medical images is needed e. g. for expert teleconsultation and remote operation planning.

Depending on the image data and the medical user different methods and systems have been developed in teleradiology, telepathology, teledermatology and teleophthalmology [1]. In many medical applications a cooperative discussion of the image contents in teleconferences is of importance enabling a common interactive view and analysis of the transferred image data.

In modern teleimaging systems for cooperative diagnostics like the teleradiology systems KAMEDIN [2] or CHILI [3] a set of image processing and visualisation functions such as zooming, edge reinforcement, level/window, contrast and brightness adjustments, marking of ROIs (Region Of Interests) etc. can be executed at the same time on the distant computers. This technique of computer supported cooperative work (CSCW) enables the distribution of an application on different computers and displays (application sharing). The image transmission as well as the communication in a cooperative teleconference are mostly based on the wide-spread communication protocol TCP/IP, whereby the application of the system becomes possible in LANs and WANs via ISDN, Ethernet, FDDI, ATM etc.. Typically, for the synchronisation of the internal messages during the teleconference a system-specific communication protocol is used which essentially consists of duplicated system instructions. Hence, in practice different teleradiology systems of different manufacturers cannot be connected in a teleconference [4]. The teleconference can be executed only if the teleradiology system is installed at both sites. The image data are transmitted beforehand, so that before the teleconference starts, the image data is available on both computers. Typically, these teleimaging systems are limited to the processing and visualisation of 2D images.

In contrast thereto, the presented teleimaging system Cypris enables the generation and cooperative visualisation of 3D models of medical image objects (tumors, bone, organs, tissues etc.) during a teleconference. Furthermore, 2D images can be transferred and discussed in cooperative teleconferences. Cypris has been implemented in Java. Therefore, the system runs on different hardware and software platforms (WINDOWS NT, WINDOWS 2000, UNIX, ...) and can easily be used in intranets and in the internet [5, 6]. In the following several characteristics of the programming language Java and the programs developed in this programming language are described.

2. JAVA

The programming language Java developed in the last years to one of the most important internet technologies. The spreading of Java applications is facilitated by the fact that Java capable internet browsers (e.g. the internet explorer, netscape navigator etc..) and Java development kits (JDK) are available free of charge. In the JDK different Java program librària called application programming interfaces (APIs) facilitate the object-oriented program development. Java can be extended by the implementation of own APIs. In the following important characteristics of the language Java are described briefly.

Java programs run on the Java virtual machine, which constructs a virtual layer between hardware, operating system and the Java program. Hence, Java programs can be executed on computers with different hardware configurations (PC, workstation etc.) and different operating systems (Windows 3.11, Windows NT, Unix etc.). The (readable) Java source code is translated by a Java compiler into (illegible) Java byte code. This generated program code can be interpreted fast by the Java interpreter. This conception facilitates dynamic loading of program code at run-time. Beyond that, run-time security is increased by the fact that Java programs are executed not directly on the underlying hardware, but on the virtual Java machine. Pre-compiled program code implemented in other programming languages can be merged by the so-called Java native interfaces.

Platform independence is a substantial prerequisite for the fact that Java programs can be executed on distant computers with unknown hardware and software configuration. Java programs running on distant computers are called applets. Java applets can be loaded from a Web servers on the local computer. At run-time further program parts can be reloaded dynamically.

Via distributed object-object communication, which is enabled by the API called remote method invocation, remote controlled method calls can be realised. This API is a very powerful tool for the implementation of distributed applications. The event handling mechanism plays an important role by the implementation of distributed Java applications. With this mechanism the information about occurring events like the pressing of a button by the user can be coded as event object and transmitted to different places of the (distributed) application.

Java based teleimaging systems can be executed platform-independently on computer with different hardware and operating systems. If the teleimaging system is realised as a Java applet, the software can be downloaded and executed even on distant computers, on which the telemedicine software is not installed. The communication partner needs only an internet browser (e.g. netscape navigator or internet explorer). Hence, a free selection of the communication partner becomes possible.

3. TELECONFERENCING IN JAVA

For teleconferencing, a new CSCW tool called Jermes has been implemented as a Java API. The name of the CSCW Tools Jermes is deduced from the words Java and Hermes. Jermes has been developed to make CSCW functions available for the cooperative use of Java programs in teleconferences application-independently.

Contrary to systems, which transfer the entire application windows or the modifications of the display contents (e.g. Netmeeting, Microsoft), Jermes synchronises the applications running on different computers on the Java event level.

By this synchronisation technique the quantity of the data communicated can be reduced significantly. This is of high importance in 3D teleimaging applications, in which the screen content changes fast. In practice, by this technique the cooperative 3D visualisation and processing of tomographic image sequences become possible on narrow-band networks such as ISDN. The Jermes API permits to make any application teleconferencable, since only Java events and no application-specific commands are transmitted.

The participating application instances exchange information about the interactions. Different communication structures are supported by Jermes. Figure 1 shows the client-server communication structure: Two computers A and B, on which client processes run, communicate over the server computer C. This architecture is suitable for the implementation of Java programs in the form of applets. With the participation of further clients it is possible to keep the state of the clients consistent by a server broadcast. As a disadvantage of this architecture high communication costs for an exchange of information between the computers A and B occur.

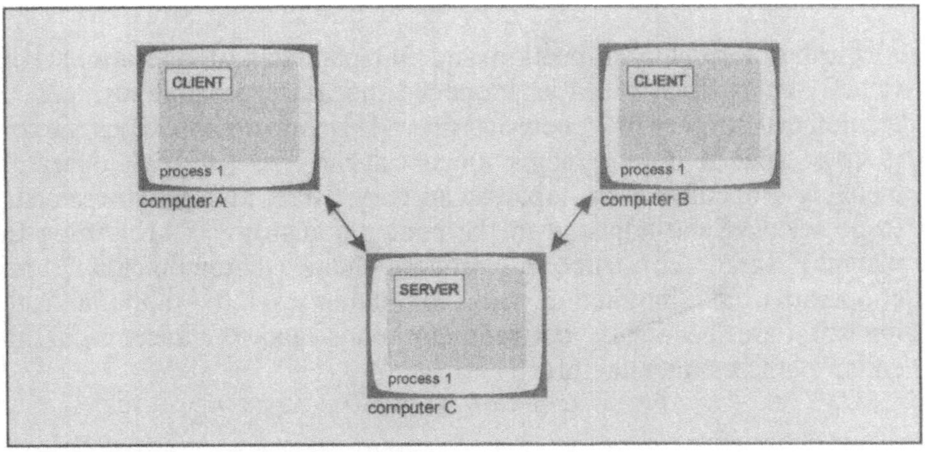

Fig. 1: Client server process communication

The implementation of the server process on a client computer avoids this disadvantage. Hence, in the implementation of the teleconference functions of Cypris the interprocess communication was realised over a thread (fig. 2).

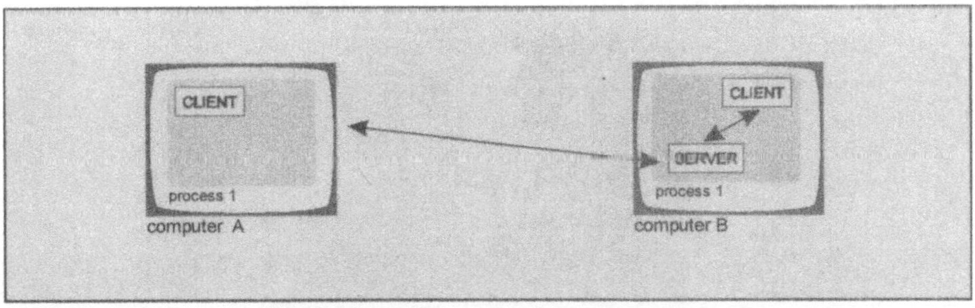

Fig. 2: Interprocess communication realised over a thread. The server process runs on a client.

4. 3D TELEIMAGING

The 3D visualisation of medial objects in threedimensional image sequences facilitates the recognition of the 3D shape and the spatial relationship between anatomical and pathological image structures. This is very helpful in the field of diagnostics, especially if the objects show complex shapes and structures like tumors or vasculary systems. In fig. 3, a 3D reconstruction of the head skin is displayed in combination with original MR images as an example. The T1-weighted MR images are mapped onto three orthogonal, user-defined planes showing a contrast-enhanced brain tumor.

Furthermore, the simulation and 3D planning of operations with virtual reality techniques has opened up new possibilities, e.g. in craniofacial surgery [7], neurosurgery [8] and orthopaedics surgery [9-14]. Special requirements occur during the virtual operation planning procedure in orthopaedic surgery. Here, endoprostheses have to be selected and adjusted to the patient's anatomy [14]. During the planning and construction process data communication and cooperative 3D interaction with 3D models of the bone and the implants are necessary between surgeons and technicians of the endoprostheses manufacturer.

The software component Cypris-3D of the teleimaging system Cypris was developed in Java to enable the cooperative 3D visualisation of medical image data. Basic functions for 3D visualisation and 3D interaction like translation, rotation and zooming of 3D objects are implemented using the Visualization Toolkits VTK [15]. Furthermore, original images can be mapped onto three orthogonal planes in the 3D scene as shown in fig. 3.

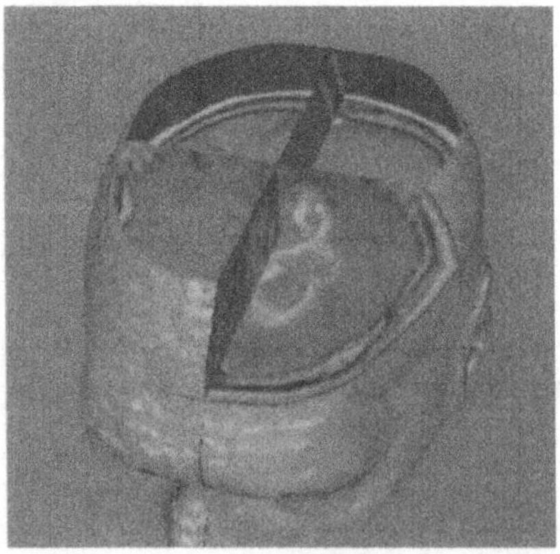

Fig. 3: 3D visualisation of the head skin. The 3D model of the skin is displayed in combination with original MR images. The T1-weigthed MR images are mapped onto three orthogonal, user-defined planes showing a contrast enhanced brain tumor.

After the integration of the CSCW tool Jermes (chapter 3) the 3D viewer of medical image data became teleconferencable. Hence, 3D reconstructions of medical image objects can be regarded and manipulated simultaneously by spatially separated teleconference users. Threshold based segmentation of medical objects can be performed [8] to define iso-surfaces and to generate 3D reconstructions of medical objects in the tomographic image sequence. Furthermore segmentation results can be imported into the system.

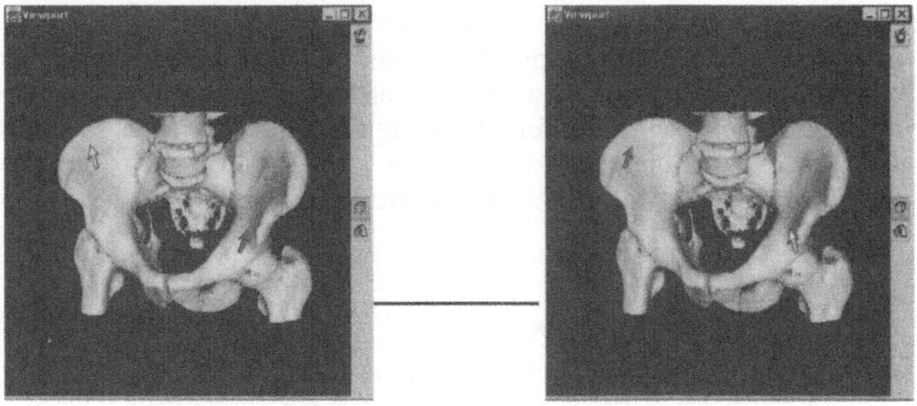

Fig. 4: 3D teleimaging session. The 3D model of the hip is displayed simultaneously on the both computer screens. During the telesession beside the local cursor the remote cursor of the communication partner is visualised (telepointing).

Before the cooperative telesession is started the tomographic image data and the 3D models of organs, bone etc. are transferred to the conference partner. Alternatively, the 3D models can be generated during the teleconference.

In practical tests, three 3D teleconferences were performed in a local area network with a mean duration of 4 minutes. The measured amount of transmitted data was less than 45 KB in each direction. Hence, with Cypris-3D telecooperative 3D teleimaging sessions can even be performed on narrow-band networks such as ISDN.

4. CONCLUSION

The developed software system Cypris enables the shared 2D and 3D visualisation of medical image sequences. During a teleconference

3D visualisations and interactions are displayed simultaneously on the screens of the teleconference partners. For the synchronisation of the telesessions the CSCW tool Jermes developed at our Institute is used. Jermes is a Java program library, in which functions for computer supported cooperative work are implemented. With Cypris 2D and 3D visualisation of medical images can be regarded together and manipulated in the cooperative teleconference. Due to the implemented communication strategy only low communication rates occur during a teleconference and even 3D teleimaging functions can be used via narrow-band networks like ISDN. In future, further features are integrated into the system to establish a Java based system for cooperative operation planning in orthopaedic surgery. In this application, for example different cutting functions as well as functions for the computer-assisted design of individually adapted endoprostheses have to be integrated into the system.

REFERENCES

[1] H. Handels, S.J. Pöppl, *Telemedizin: Grundlagen-Perspektiven-Systeme-Anwendungen*. Aachen: Shaker, 1999.

[2] H. Handels, C. Busch, J. Encarnacao, C. Hahn, V. Kühn, J. Miehe, S. J. Pöppl, E. Rinast, C. Roßmanith, F. Seibert, and A. Will, "KAMEDIN: A Telemedicine System for Computer Supported Cooperative Work and Remote Image Analysis in Radiology," *Computer Methods and Programs in Biomedicine*, vol. 52, pp. 175-183, 1997.

[3] U. Engelmann, A. Schröter, U. Baur, M. Schwab, O. Werner, M. H. Makabe, and H. P. Meinzer, "Second Generation Teleradiology," in *CAR 98- Computer Assisted Radiology*, H. U. Lemke, M. W. Vannier, K. Inamura, and A. Farman, Eds. Amsterdam: Elsevier, 1997, pp. 437-442.

[4] H. Handels, E. Rinast, C. Busch, C. Hahn, V. Kühn, J. Miehe, C. Roßmanith, F. Seibert, and A. Will, "Image Transfer and Computer-Supported Cooperative Diagnosis," *Journal of Telemedicine and Telecare*, vol. 3, pp. 103-107, 1997.

[5] K. Kleber, A. Schröter, J. Hostein, and D. H. W. Grönemeyer, "Jive - Ein Konzept zum Java-basierten Zugriff auf DICOM-Archive," in *Telemedizin: Grundlagen-Perspektiven-Systeme-Anwendungen*, H. Handels and S. J. Pöppl, Eds. Aachen: Shaker, 1999, pp. 18-24.

[6] H. C. Klaiber, H. Handels, I. Grande-Nagel, U. Knopp, and S. J. Pöppl, "Java-basierte Teleradiologie - Neue Möglichkeiten und Anwendungsszenarien.," in *Telemedizin: Grundlagen-Perspektiven-Systeme-Anwendungen*, H. Handels and S. J. Pöppl, Eds. Aachen: Shaker, 1999, pp. 25-30.

[7] E. Keeve, S. Girod, and B. Girod, "Craniofacial Surgery Simulation," in *Visualization in Biomedical Computing*, vol. 1131, *Lecture Notes in Computer Science*, K.-H. Höhne and R. Kikinis, Eds. Berlin: Springer Verlag, 1996.

[8] R. Shahidi, "Applications of Virtual Reality in Stereotactic Surgery: Volumetric Image Navigation via Surgical Microscope," in *Computer Assisted Radiology*, H. Lemke, U., K. Inamura, C. C. Jaffe, and M. W. Vannier, Eds. Berlin: Springer Verlag, 1995, pp. 1126-1138.

[9] J. E. Moody, A. M. DiGioia, B. Jaramaz, M. Blackwell, B. Colgan, and C. Nikou, "Gauging Clinical Practice: Surgical Navigation for Total Hip Replacement," in *Medical Image Computing and Computer-Assisted Intervention - MICCAI 98*, W. M. Wells, A. Colchester, and S. Delp, Eds. Berlin: Springer Verlag, 1998, pp. 421-430.

[10] J. A. Richolt, M. Teschner, P. Everett, B. Girod, M. B. Millis, and R. Kikinis, "Planning and Evaluation of Reorienting Osteotomis of the Proximal Femur in Cases of SCFE Using Virtual Three-Dimensional Models," in *Medical Image Computing and Computer-Assisted Intervention - MICCAI 98*, W. M. Wells, A. Colchester, and S. Delp, Eds. Berlin: Springer Verlag, 1998, pp. 1-8.

[11] C. J. Sutherland, S. J. Bresina, and D. E. Gayou, "Use of General Purpose Mechanical Computer Assisted Engineering Software in Orthopaedic Surgical Planning: Advantages and Limitations," *Computerized Medical Imaging and Graphics*, vol. 18, pp. 435-442, 1994.

[12] D. A. Simon, B. Jaramaz, and M. Blackwell, "Development and Validation of a Navigational Guidance System for Acetabular Implant Placement," in *First Joint Conference of CVRMed and MRCAS*. Grenoble, France: Springer Verlag, 1997, pp. 583-592.

[13] C. Y. Tso, R. E. Ellis, J. Rudan, and M. M. Harrison, "A Surgical Planning and Guidance System for High Tibial Osteotomies," in *Medical Image Computing and Computer-Assisted Intervention - MICCAI 98*, W. M. Wells, A. Colchester, and S. Delp, Eds. Berlin: Springer Verlag, 1998, pp. 39-50.

[14] H. Handels, J. Ehrhardt, W. Plötz, and S. J. Pöppl, "Virtual Planning of Hip Operations and Individual Adaption of Endoprostheses in Orthopaedic Surgery," *International Journal of Medical Informatics*, vol. 58-59, pp. 21-28, 2000.

[15] W. Schroeder, K. Martin, and B. Lorensen, *The Visualization Toolkit - An Object-Oriented Approach To 3D Graphics*. New Jersey: Prentice Hall, 1998.

3D Image Based Collaboration System for Telemedicine

[1]TOSHIHIKO KOBAYASHI, SYUJI SATOU, [2]HAO JIANG, TETSUYA FUJII, [3]NOBUO SUGOU, TOSHIAKI MITOU AND IEKADO SHIBATA
[1]KGT Inc., Shinjuku-ku, Tokyo, Japan: [2]Communications Research Laboratory, MPT, Koganei-city, Tokyo, Japan: [3]Department of Neurosurgery, Toho University School of Medicine, Tokyo, Japan

Key words: 3D Medical image, Volume rendering, Collaboration, Avatar, Telemedicine

Abstract: This paper describes a 3D image based collaboration system for telemedicine. This system enables doctor to observe 3D images transferred from image database through network, and also has collaboration function which operate 3D position of viewing of each doctor by 3D image displaying and data transferring through network. It is constructed based on server client style, and the server has the function of transferring data from a 3D image database and a collaboration record database, and also has control unit of transferring data between doctors during operating the system. The client has a user interface including operation part for parameter selection of viewing and sending comments, and display part for displaying 3D image based on volume rendering and 3D position and direction of observer by using avatar. Doctors can use this system to do collaboration work by sending comments with viewing 3D image each other. We implement this system in different platform including UNIX workstation and PC, and also supply Web browser based user interface for considering various user environments. We applied the 3D images of MRI in head for examining the structure of soft tissues and tumor in detail. We also evaluated the performance of the system through network including LAN and Internet, and experimental result shows that this system is useful.

1. INTRODUCTION

As progress of 3D medical device, X-ray CT and MRI have been widely applied in hospitals for examination and disease diagnosis, so that 3D image based diagnosis is very important method for doctor to decide disease and plan surgical operation[1-2]. On the other hand, collaborative environment construction based on network and computer technologies is a very important way for providing high quality of service[3], especially how deal with large and high dimension data of medical image, such as X-ray CT and MR[4]. From multiple points can give an impact to increasing the quality of medical service. Several studies have been reported, however, most of all have been based on exchanging 2D images and 2D movies[5-7].

We have been developing a 3D image based collaboration system for medical application. This system can operate 3D medical images and 3D position of viewing of each doctor by 3D image displaying and data transferring through network. It is constructed based on server client style, and the server has the function of transferring data from a 3D image database and a collaboration record database, and also has control unit of transferring data between doctors during operating the system. The client has a user interface including operation part for parameter selection of viewing and sending comments, and display part for displaying 3D image based on volume rendering and 3D position and direction of observer by using avatar. Doctors can use this system to do collaboration work by sending comments with viewing 3D image each other.

In this paper, we describe the design of the 3D medical image based collaboration system and show the experimental results.

2. SYSTEM CONCEPT

To design the system, we have several concepts. First, all images for displaying are of 3D rendering, and all operations on the images are curried out three-dimensionally. Second, users can be exchanged multi-modal information including text and image during collaboration. Final, the system can be used in multiple terminal environments from PC to workstation.

2.1 3D image based observation

All images for observation or collaboration are the 3D image based on rendering. The comments exchanged between users and the position and direction of user during operation are also displayed three dimensionly. In the other words, we construct a virtual space and display the 3D image, so that user can access the space through network, and have the same space for collaboration.

2.2 Multi-modal information exchange

During collaboration, the images of viewing are needed for each user. Also, such text of other information is also important for communication between users. The system allows user to input comments into the 3D image displayed in the virtual space, directly and three dimensionally. It can increase the performance among users.

2.3 Multi-terminal use

Since computer and network environment is different for each user, we implemented this system in different platform, including UNIX workstation and Windows PC to draw the best performance based on each platform.

3. SYSTEM OVERVIEW

3.1 Outline

Figure 1 shows the system overview. We applied the 3D images of MRI in head for examining the structure of soft tissues and tumor in detail. The system consists of three parts, user interface, database server, and database. We have two databases of 3D image database and collaboration record database. The user interface consists of three parts, database controller, viewer, and collaboration user interface.

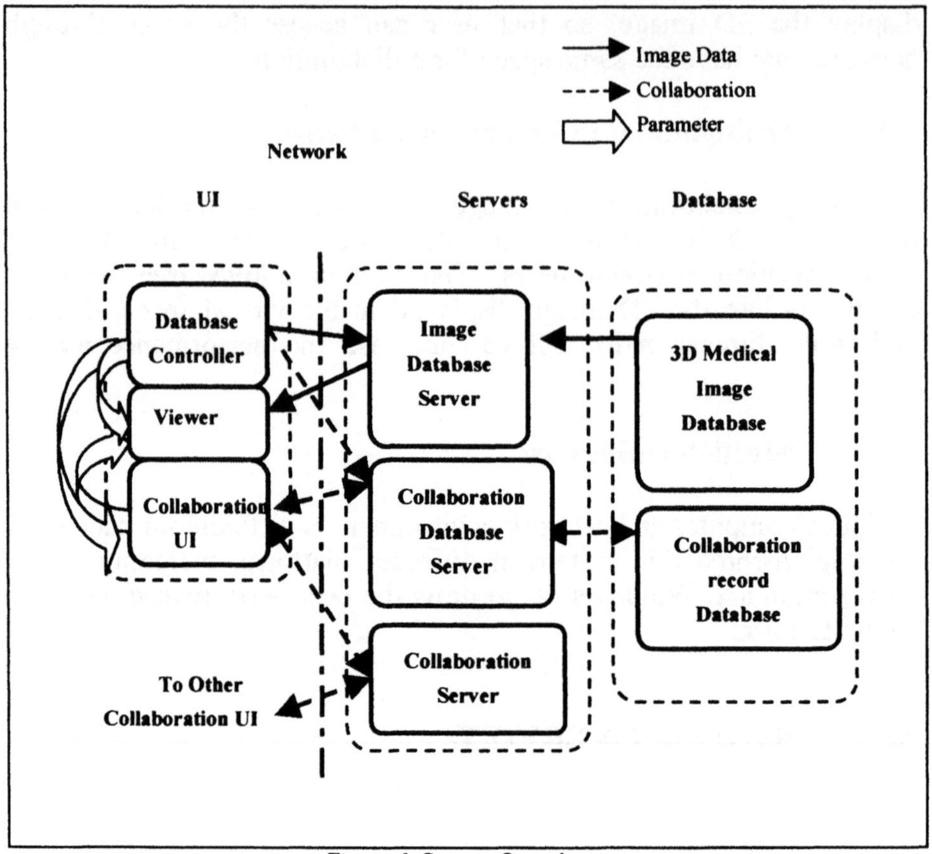

Figure 1. System Overview.

3.2 Database and Server

We applied 3D image data and collaboration record onto Object-Oriented database structure. The 3D image database stores 1) volume data obtained from CT and MRI (256x256x256 16bit), 2) surface data of each tissue in the brain with VRML1 format, and 3) text information on patient, tissues etc.

The image database server sends the image data after received the request from client. Collaboration database server stores and delivers the record of collaboration, and the collaboration server exchange the parameter data among clients for viewing during collaboration at the same time.

3.3 User Interface

User interface consists of three parts, database controller, viewer, and collaboration user interface. The database controller allows user to browse the contents in the database by tree structure and catch the data information with thumbnail image, then user can send a request to the database server to obtain image data for viewing. The tree structure is implemented with Java so that the database controller obtains the data structure from database server at once. User can select a part based on the structure, which is needed for observation, and send a request to database server for transferring the real image data. The viewer displays the 3D image based on the request of controller after receive the real image data from databases. The outline among database, database controller and viewer is shown in Figure 2.

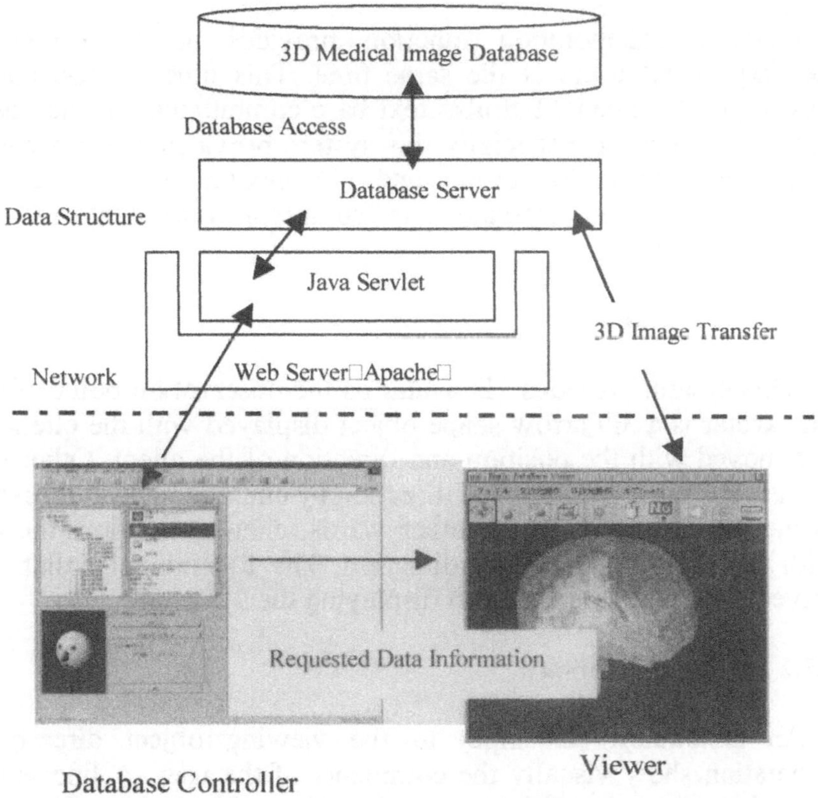

Figure 2. The outline among 3D image database, database controller, and viewer

4. 3D COLLABORATION FUNCTION

This system has two functions to support collaboration work among doctors, on time collaboration and off time collaboration, we called. On-time collaboration function provides a meeting environment for all doctors at the same time, and all comments and parameters for viewing of doctors are recorded as a data set of conference proceeding in the collaboration record database. Off-time collaboration function allows doctor to refer the past conference and add comments by displaying the 3D image based on the recorded data set. On the other words, this function can reproduce the past conference any time.

4.1 On-time collaboration

On-time collaboration function provides an environment of meeting for all users at the same time. This function has not only viewing of 3D image, but also text base communication such as chat and 3D annotation. Especially this system provides client a common observation space. 3D avatar and 3D annotation are used in the common space to increase the communication performance of collaboration work.

4.1.1 3D avatar[8]

This system provides 3D avatar as the observation point of client. The avatar is a 3D arrow shape object displayed with the client name and moved with the position and direction of the client. Other clients can get the same position and direction by clicking an avatar displayed in the virtual space. In the other wards, clients can view the object with the same position and direction. The function of collaboration server is to send parameters of displaying the 3D avatar.

4.1.2 3D annotation

An annotation can input to the viewing object directly. The annotation show visually the comments of the user. A line is drawn from the view point of the user to the object of interest, and comments is shown in character forward the user always. Figure 3 shows a sample image for on time collaboration between two clients. The left image of the is the user named "Nobuko", and the right image is the user "Satou". "Nobuko" is observing the 3D image from left side by sixty degree of "Satou" 's. Each view has an avatar of another person with the name and 3D annotation on the 3D image. 3D annotation can point out and provide information independent of viewing direction of observation.

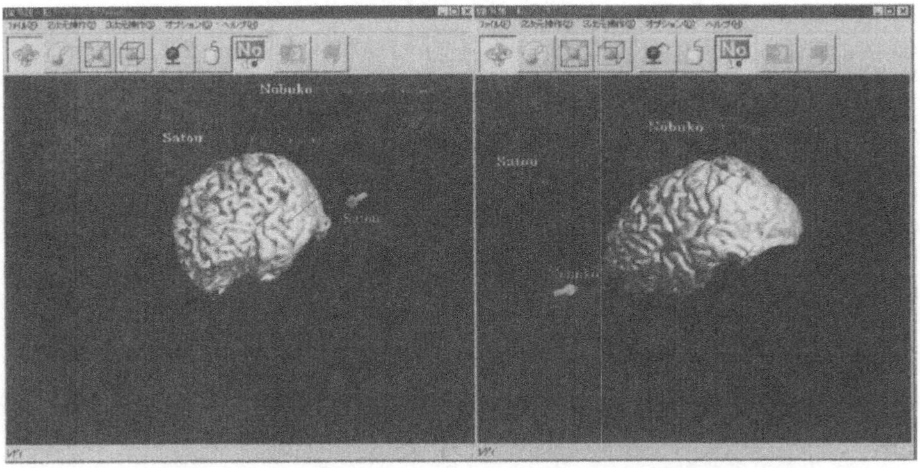

Figure 3. The sample image of two collaborating client viewer

4.2 Off-time collaboration

Because of time schedule, user can attend the past conference by using off-time collaboration function. The layer structure of off-time collaboration database shown in Figure 4. This GUI allows user not only viewing the past collaboration result but participate in the collaboration with adding a new comment under an "meeting room".

A "house" type icon represents a meeting room and an "arrow" type icon represents titles of the interim reports. If the participant of the off-time collaboration selects an icon, he/she receives procedure of the meeting and comments from collaboration database and the images from 3D image database, then reproduce the past collaboration in viewer. Figure 5 shows a starting state of the off-time collaboration. Of course, the functions of on-time collaborations are available in the off-time collaboration.

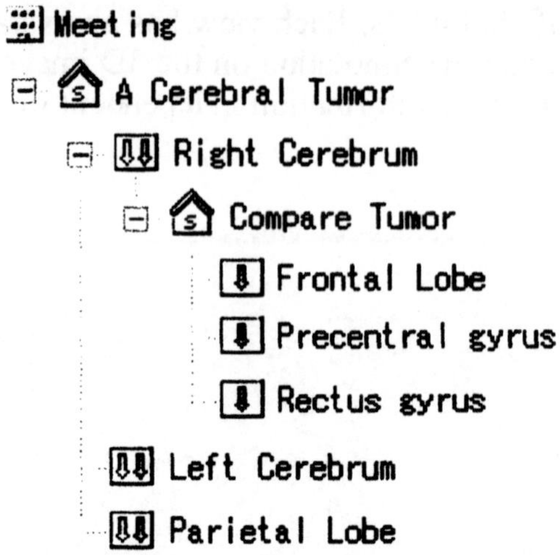

Figure 4. Off time collaboration GUI

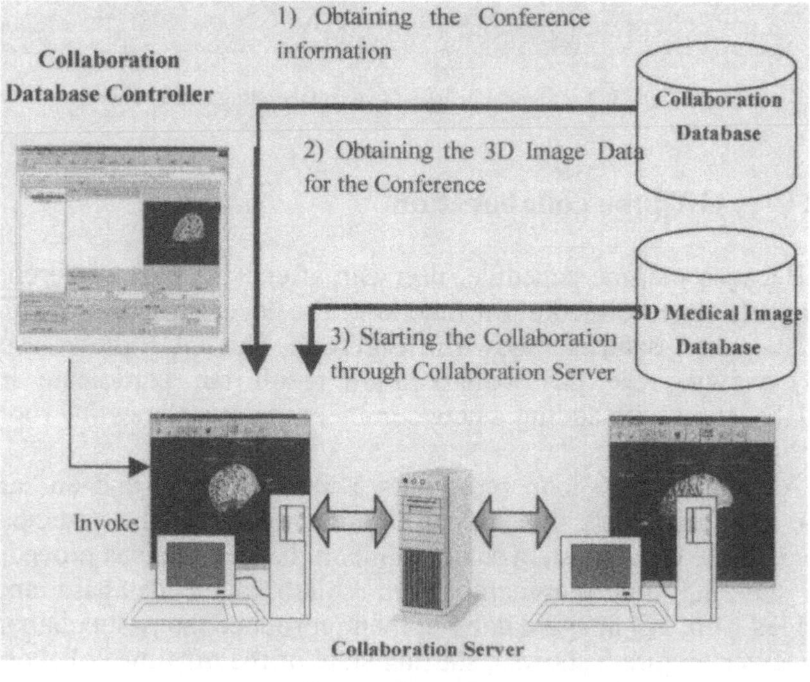

Figure 5. The first state of off-time collaboration

5. EXPERIMENTS

This experiment shows that how long time is needed to reflect in to other client if one client changed the data of displaying the 3D avatar. We prepared a MRI volume data (256x256x256), the size of the volume was almost 2MB after zlib compression, and stored them into Object-Oriented database running on SGI Octane (IRIX6.5). Three

Windows PC clients were connected to database server and transferred the volume data by TCP/IP socket then rendered with volume rendering. One client was connected via LAN (10baseT) and other two clients were connected via Internet with 64K ISDN(Integrated Service of Digital Network). The parameter data of displaying 3D avatar was exchanged by collaboration server using TCP/IP socket and reflected other clients. The benchmark was to measure the throughput from starting avatar event and receiving it through collaboration server.

The data transferring of avatar is small enough to communicate through Internet, so the time to exchange avatar data is 0.5 second (LAN) and 1.6 second (ISDN) on average. We can recognise the 3D annotation and avatar assisting us smooth collaboration work.

6. CONCLUSION

We have been developing a 3D image based collaboration system. This system has three features, 3D image-based observation, multi-modal information exchange, and multi-terminal use. The system consists of three parts, user interface, database server, and database. For collaboration, we designed two functions "on-time collaboration" and "off-time collaboration". On-time collaboration provides 3D avatar and 3D annotation to carry out smoothly collaboration work. The conference record can be stored in database including comments, images, and operation procedure, so that users can review any conference from the database in any time. We applied 3D medical image data to evaluate for this system. The test result for 3D avatar and 3D annotation shows the efficiency of these functions for assisting 3D image based collaboration work on the network. We recognized this system is efficient under various computer and network environment.

REFERENCES

1. Y. Itai, K. Katada, N. Kobayashi, 1996, *Japanese Journal of Clinical Radiology 41*, Kinbara.

2. This is Medicine in Japan, 2000, ME, pp.145-153.

3. K. Hosoya, Y. Kato, A. Kawanobe, S. Kakuta, and Y. Fukuhara, , 1997, *Collaborative Educational Environment Based on Multi-User Virtual Space*, IEICE, Vol.J80-D-II, No.4, pp.866-873.

4. I. Pyarali, 1996, *Design and Performance of an Object-Oriented Framework for High-Speed Electronic Medical Imaging* Computing Systems Journal, USENIX, Vol. 9, No. 3.

5. *Medical Collaboratory Testbed*, published on the WWW, http://www.si.umich.edu/medcollab/index.html

6. *Dx Multimodality*, published on the WWW, http://medasys.co.jp

7. T. Matsuda, *A STUDY ON AUTOMATIC RECONSTRUCTION OF THE 3D SOLID SKULL MODEL USING COMPUTER TOMOGRAPHY DATA*, Published on the WWW,http://pegasus.sfc.keio.ac.jp/~tatsuki/master/Main.html

8. *Virtual Society*, published on the WWW, http://vs.sony.co.jp/.

Computer Mediated Communication of Medical Image Data: An Example from Cardiological Ultrasound Data

[1]CHRISTOPH WICK, [1]STEFAN PIEPER AND [2]MICHAEL WEIDENBACH
[1]*GMD National Research Center for Information Technology, HCI Research Department, Schloß Birlinghoven, D-53754 Sankt Augustin, Germany*
[2]Department of Pediatric Cardiology, University Bonn, Adenauerallee 119, D-53113 Bonn, Germany, *Email: christoph.wick@gmd.de*

Key words: Computer mediated communication, teleconsultation, 3D ultrasound, virtual reality

Abstract: This paper dicusses a system that supports telecommunication of 3D cardiological ultrasound data sets of the human heart. The system enables the participating physicians to select arbitrary 2D views from a 3D ultrasound data set and simultaneously visualize additional information such as view orientation and "regions of interest" using a virtual heart model. Image and model can be annotated and stored together with view parameters as a "case description" in a central database for later access or be sent as a "multimedia Email" to a consultation partner.

1. INTRODUCTION

Image communication plays an increasing role in networked medicine. Generalists consult specialists for a second opinion. Training and education requires prepared material for teaching. Quality control forces physicians to write down examinations for later use.

For almost 15 years PACS (Picture Archiving and Communication Systems, see [2]) claim not only to "archive" (in terms of "storing") but also to "communicate" medical images. Recent implementations of

PACS, such as various DICOM systems (see [8]), interpret the term "communication" in a very technical way, namely transportation of data and ubiquitous access to these data.

For communication from a human point of view, transport of data, especially image data, is not communication in a narrower sense of the definition. Actually not only the data shall be transported from a sending physician to a recipient but his findings or - to speak in other words - his image *interpretation.*

DICOM Structured Reporting (see [4]) is an approach to also integrate findings and annotations into the DICOM framework, but until today it has not been widely adopted.

In the field of CSCW (Computer supported cooperative work, see [1]), considerable research about how computers can support distributed work has been performed. Unfortunately most of the work has been targeted towards the field of office applications such as word processing. In this case the artifact to work on (e.g. a text document) is basis, context and piece of work in one glue.

Communication of findings based on medical image data, especially the field of three-dimensional echocardiography, is a more complex task. Here we must deal with a complex relation of data interpretation (e.g. images), annotation of these data and their contextualization into a three dimensional context (e.g. anatomical structures).

2. RELATED WORK

Three-dimensional echocardiography (see [5]) offers the advantage of acquiring and storing 3D volume data sets that cover the complete heart. The data sets can be stored into a (remotely accessible) database for later examination using postprocessing software such as the 3DE viewer, described later.

3DE offers advantages over conventional 2D echocardiography:

- Once acquired and stored into a database, examinations can be performed not only when the patient is present, but in any place and at any time where the viewer and the data set are available.

- Since the physician can navigate freely through the 3DE data set, a physician responsible for the second opinion is not forced to rely on his colleague's pictures. He can also navigate through "regions" of the data set others have not analyzed since they are still accessible through the nature of the 3DE data set.

At GMD's HCI Department we have previously developed a viewer for 3D echocardiographic data sets (3DE) (see [11], [10]). This 3DE viewer uses slicing as the visualization technique; it computes plane 2D images from a 3D volume data set and displays them planar on the screen, similar to the conventional 2D ultrasound images a trained physician is familiar with. position and orientation of the slice in relation to the data set can easily be controlled by either selecting predefined standard positions (see [7]) or using 3D input controls for free hand navigation.

Since the viewer allows the examining physician to navigate freely through the ultrasound data, each data set is registered with a virtual heart model. Besides displaying the slice image itself, the viewer offers a three dimensional view of model and a virtual representation of the slice plane indicating position and orientation of the slice in relation to the patient's heart. To reinforce understanding of the anatomical structure in the slice image the image can be overlaid with contours taken from the model.

By this means the virtual heart model provides visual guidance to the physician and avoids a "lost in space" effect. The model enables him to interpret the slice images correctly.

3. MOTIVATION AND GOALS

Medicine has a long tradition in transcribing to paper examinations and findings. Codesystems, such as MeSH (see [9]), or Snomed (see [3]), provide a standardized basis for documentation. "Annotation" of images can be done easily too - if necessary - by simply drawing on them.

The more findings rely on digital images the more a precise replication of one communication partner's displayed information becomes necessary. The way images appear on a physician's display does not depend only on the digital data stored in a PACS but also on the parameters with which the appendant viewer is parameterized, e.g. brightness, contrast or color depth.

In the case of 3DE data sets it is even more complicated. Simply sending a raw data set to a communication partner with a textual annotation like "muscular ventricular septal defect" is quite useless since the recipient may be unable to verify this diagnosis due to lack of precise knowledge of the position and the orientation of the inspected slice image. He must arduously try to reselect the originally inspected slice plane - a tedious and error prone task.

Graphical annotation of the data set is not possible either because the current 3DE viewer offers no means to attach graphical annotations to a selected slice image.

Printing out (or making a digital screenshot of) the selected slice image removes the advantage of inspecting other regions of the data set freely. The benefits of 3DE would be lost. A recipient must rely on the slices the originator had selected.

Therefore the above described viewer must be extended for communication features. It should offer possibilities to:

- store a selected slice position for further examination,
- annotate a selected slice image with annotational drawings,
- attach annotations to the heart model to clarify findings in the slice image,
- embed this annotations into a textual description, and
- file an annotated data set for later retrieval or pack and send it as a multimedia Email to a communication partner.

4. SYSTEM DESCRIPTION

We have described requirements to make a communication tool from the 3DE viewer. Techniques shall be suggested which address these requirements.

4.1 Slice plane bookmarks

As described above the 3DE viewer offers two ways to select slice images from the 3DE data set in two ways: either by selecting a predefined standard position from a menu or by freely selecting a slice image using 3D input controls. To solve the problem of preserving freely selected slice images, a mechanism has been added to the 3DE viewer similar to bookmarks in web browsers. First the physician selects a position (like a web page), then this position can be made persistent, named and stored for later retrieval (comparable to the "add bookmark" function). This slice image is now accessible easily through the menu using the same technique as accessing predefined standard positions. Figure 1 shows a screenshot of the system.

4.2 Slice image annotation

The problem of image annotation can be solved by adding pointing or surrounding drawing tools overlaying the slice image. Therefore the physician chooses a drawing tool and marks the region of interest by dragging the cursor. To gain access to the unobscured image these drawings can be hidden interactively.

The annotation of the slice images are related to a single selected slice image. They are fixed in the "2D space" of the plain images. If the user selects another slice image the annotations disappear.

Figure 2 (left side) shows a screenshot of an activated annotation around the mitral walve.

4.3 Model annotations

The annotation of the slice images are related to a single selected slice image. They are fixed into the "2D space" of the plain images. If the user selects another slice image the annotations disappear.

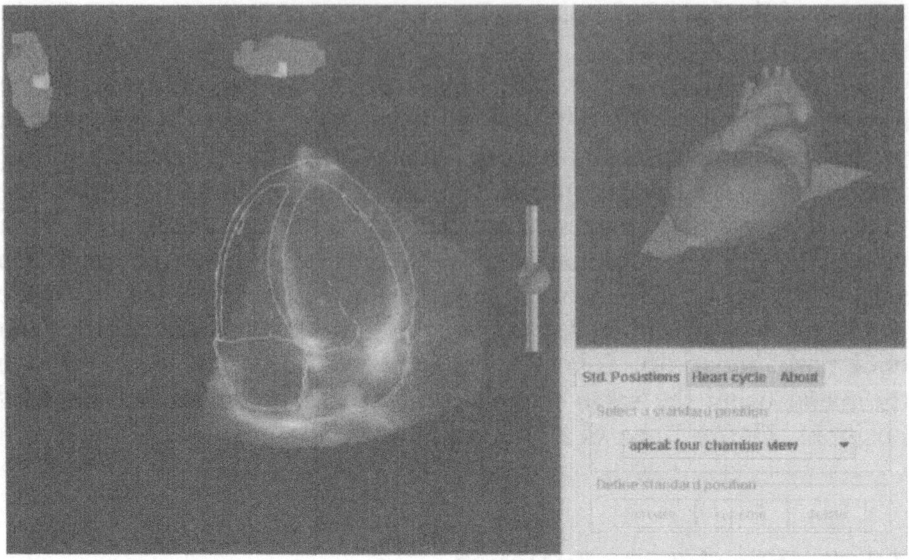

Figure 1: Screenshot of the system with a selected standard position. Below the standard position menu the buttons for creating, renaming and deleting "bookmarks" are visible.

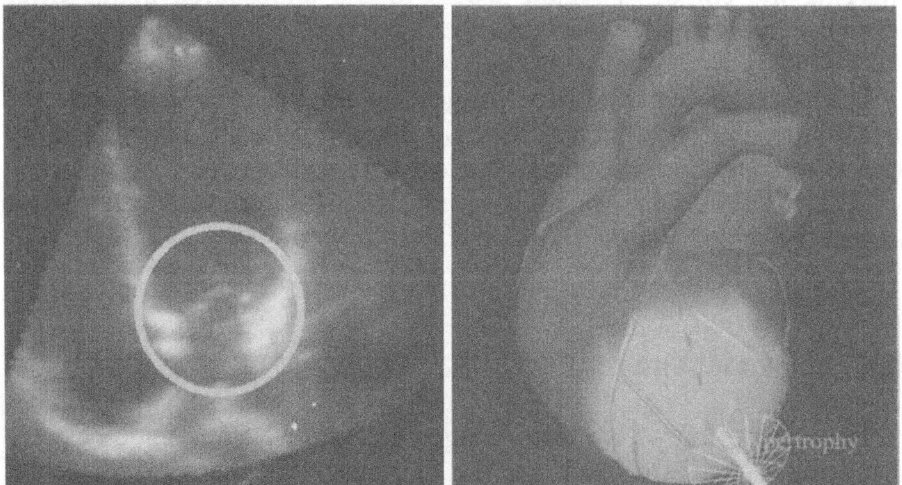

Figure 2: **left**: Screenshot of a slice image with an annotation around the mitral valve. **right** Screenshot of an annotation of the heart applying a hypertrophy onto the ventricle.

To embed an annotation into its three dimensional context the heart model can be annotated too, using tools described in [6]. They modify the heart model in a way that anatomical structures are highlighted or pathologies are emphasized. These modifications on the heart model can be made visible in all slice image transformations that are "cutting" the annotation using the contours overlay.

Figure 2 (right side) shows a screenshot of a model, annotated with a hypertrophy.

4.4 Textual annotations

Every slice image of interest, with or without annotations, must be embedded into a verbal description, which may transport findings from the originator to the recipient. This verbal descriptions transposes *data* into *information*.

A communication tool therefore must provide an editor to transcribe findings and diagnoses. To link the image-based findings into a textual description, we suggest to insert small icons into free text. These may act as *hyperlinks* to selected positions or annotations of slice images or the model. If the user clicks them the 3DE viewer displays the related slice image or model setting immediately. By the help of a context menu the editor can also expand the hyperlinks "on the fly" and embed the linked media directly into the free text. This gives the recipient the choice of a compact vs. detailed presentation.

4.5 Pack and go

All texts, annotations and systems settings can be stored in a central, https accessible database based on user identification, a time stamp and a version identifier. Each contributing editor can adjust access control on three steps: private, access to restricted users (e.g. for second opinion), and common access (e.g. for training and documentation).

He can notify communication partners via Email about newly annotated cases. The recipient will then receive a short Email with a hyperlink which will pop up an enhanced 3DE viewer and parameterize it with the respective case description.

To grant access to communication partners who do not have access to the central database or to prepare for offline examination, a whole case description (data set plus annotations) can be exported into one single file which can be sent as multimedia Email.

5. DISCUSSION

The system targets the combination of several tasks under a uniform user interface. Firstly, through the combination of slice image and heart model, it provides awareness of the spatial orientation and transformation of the slice plane in relation to the anatomy of the human heart. Secondly, not only single slice images itself, but also the heart and - through the resulting contours - multiple images can be annotated by the user. This kind of 3D annotation supports the congruency of spatial classification of the findings for both communication partners. The annotated model acts as a common artifact (see [1]) which provides visual guidance through the 3D space as well as for the perception of the communication partners.

The integration with textual annotations is solved in an unobtrusive and adjustable manner. It preserves the advantages of 3DE data sets with an easy to read and modify interface for case descriptions. This combination of data set and textual annotation especially distinguishes it from conventional PACS.

6. CONCLUSION AND FURTHER RESEARCH

The suggested improvements for a 3DE viewer provide the possibility to communicate findings from an examining physician to a recipient. Although a formal evaluation of the user interface is pending, first feedback from physicians has shown that our solution is viable and is widely accepted by its intended user base.

Further research will be conducted the area of integration of the system into other PACS systems and the integration into publicly accessible web based training systems.

REFERENCES

[1] Berlage, T., Sohlenkamp, M.: Visualizing common artifacts to support awareness in computer-mediated cooperation, 1999, Computer supported cooperative work ; Vol. 8, Nr. 3

[2] Bick, U., Lenzen, H.: PACS the silent revolution. European Radiology, Volume 9, Issue 6, pp 1152-1160 .

[3] Bidgood, Jr. W.D.: The SNOMED DICOM Microglossary - Controlled, Terminology Resource for Data Interchance, in Biomedical Imaging, Meth Inf Med, 1998, ;37, pp. 404-414

[4] Bidgood D, DICOM Working Group 8: Digital Imaging and Communications in Medicine (DICOM) Supplement 23: Structured Reporting (Draft). Northfield, IL: College of American Pathologists, 1999.

[5] Bruining, Nico; Roelandt, Jos R.T.C.; Grunst, Gernoth; Berlage, Thomas; Waldinger, Johannes; Mumm, Bernhard: Three-Dimensional Echocardiography: The Gateway to Virtual Reality, Echocardiography, Bd. 16, Nr. 5

[6] Fox, T., Weidenbach, M. (1997): TiDAS - Interaktive dreidimensionale Gestaltung von angeborenen und erworbenen Herzfehlern. In: Baur, M.P.; Fimmers, R. & Blettner, M. (Hrsg.): Medizinische Informatik, Biometrie und Epidemiologie - GMDS '96, 41. Jahrestagung der GMDS Bonn, September 1996. München: MMV Verlag, 187-190.

[7] Henry, W. L., DeMaria, A.,Gramiak, R.: Report of the American Society of Echocardiography committee on nomenclature and standards in two-dimensional echocardiography. Circulation 2:212-217.

[8] National Electrical Manufacturers Association (NEMA): Digital Images And Communications in Medicine (DICOM), NEMA Standard PS 3.1-1999 - PS 3.14-1999, http://www.nema.org

[9] National Libray of Medicine (NLM): Medical Subject Headings (MeSH), http://www.nlm.nih.gov/mesh/meshhome.html

[10] Weidenbach, Michael; Wick, Christoph; Pieper, Stefan; Quast, Klaus-J• urgen; Fox, Thorsten; Grunst, Gernoth; Redel, Dierk A.: Augmented Reality Simulator for Training Two-Dimensional Echocardiography, COMPUTERS IN BIOMEDICAL RESEARCH, Volume. 33, Issue. 1, pp. 11-22

[11] Wick, Christoph; Pieper, Stefan; Weidenbach, Michael; Redel, Dierk A.; Berlage, Thomas: First experiences with cardiological teleconsultation using 3D ultrasound data sets, in Lemke Heinz U.: CARS'99 - Computer Assisted Radiology and Surgery, 1999, pp. 61-66.

Digital Image Communication in Radiotherapy

From Proprietary Formats to a DICOM Based Network

WOLFGANG W. BAUS AND BERND KOBER
Institut für Radioonkologie/Strahlentherapie, Klinikum Darmstadt, Grafenstr. 9, Darmstadt, Germany

Key words: RT Structure Set; RT Plan; RT Dose; RT Image; RT Treatment Record.

Abstract: The specific needs of radiotherapy in respect to image communication are illustrated by presenting the modalities involved in the treatment process. The existing network solutions explain the demand for a general standard. This standard will be DICOM RT. The impact on radiotherapy networking and the present status of the implementation will be discussed. Considerations on the possibility of teleradiotherapy are also included.

1. INTRODUCTION

In radiotherapy (RT), diagnostic images usually form the basis of the therapy decision. However, during the treatment planning process image information is mingled with other data which is what makes therapy image communication different from other fields. It is also the reason why standardization and digital image communication lacks behind diagnostic radiology. This is not in contradiction to the fact that a state of the art therapeutic department shows a high level of digital networking. One limitation for communication is the lack of an overall standard. Instead, only lots of different proprietary interfaces are available. For every new investment the compatibility with the existing communication chain has to be ensured. Usually afterwards a lot of 'exceptions' appear where the digital information has to be corrected manually after transmission. There is clearly a need for an overall standard which includes both parameter and image communication.

This standard will be the RT-extension of the DICOM[i] standard which will be a main topic of this paper. Maybe with the help of the standard Teleradiotherapy – which is not an issue in radiotherapy to date – could then also come on the agenda.

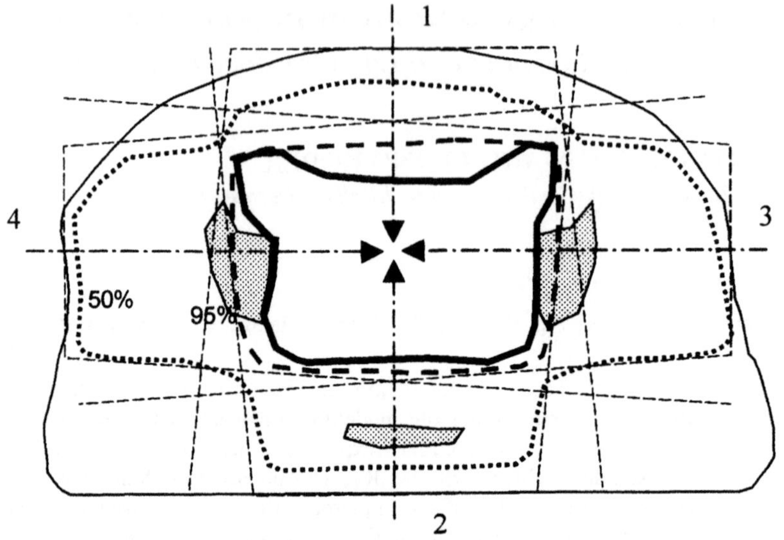

Figure 1. Transversal slice of a patient in supine position derived from a CT-scan (omitted here) of the pelvic region with important structures: the patient body, the bones (grey areas) and the target volume (bold line). Already included is the beam arrangement (four-field box plan, arrows and dashed lines) and the calculated isodoses (dashed: 95%, dotted: 50%).

1.1 Radiation Therapy

Radiotherapy means in most cases the application of photons or electrons produced by linear electron accelerators (LINAC, external beam therapy).[ii] The gantry with the beam exit can be rotated in a plane transversal to the patient. If the tumour is localized around the axis of rotation of the LINAC and different angles are used then the dose can be concentrated around the tumour merely by geometry. This leads to some typical set-ups of which the so-called box for treatment of the pelvic region may serve as a simple example (Fig. 1). The planing process starts with a transversal CT-slice from which the anatomical

i DICOM stands for Digital Imaging and Communications in Medicine.

ii The demands in respect to image communication of other methods (e.g. brachytherapy, hadrontherapy) are in principle the same as in external beam therapy and therefore not considered here.

contours are derived (body, bones). This is necessary for inhomogeneity corrections during dose calculation or – in case of organs at risk – for a so-called dose-volume histogram. The physician then delineates the tumour area plus safety margin and lymph nodes according to the extension of the disease (thick line). The CT-scan together with the structures and markers is then transferred to the treatment planning system (TPS) where the medical physicist chooses the beam angles, field sizes and possibly beam modifiers (blocks, wedges, bolus, etc.). This treatment plan is the basis of the dose calculation of the TPS. Usually, the dose is visualized as lines of equal dose (isodose plan, dashed and dotted line in fig. 1). The accepted treatment plan is checked with the help of a machine which equals a LINAC in many respects, except that the radiation source is replaced by a diagnostic x-ray tube (Simulator). These simulation images (or comparable digitally reconstructed radiographs, DRR) which always correspond to a beam's eye view (BEV) are necessary for documentation of the treatment (intention) and for verification during application.

The plan is now fed into the record - and verify system (R&V) which forwards the treatment parameters to the LINAC and keeps track of the course of the treatment which usually stretches over a multitude of sessions.[iii] The R&V system is one guarantee for the same set-up at each treatment day. The other is verification of the patient position by image modalities such as Electronic Portal Imaging Devices (EPID).[iv] By matching these images with the simulator images or DRRs the physician checks the daily set-up. In radiotherapy, image information is almost vitally connected with other data or parameters, as will be discussed later in more detail.

2. THE RADIOTHERAPY NETWORK

All the modalities mentioned in the previous section (CT, TPS, Simulator, R&V, LINAC, and EPID) are more or less digitally connected (or, at least they can be connected) in a modern state of the

iii A typical total dose of a treatment is 50 Gy. Usually, it is applied in fractions of 2 Gy per day because tumours and normal tissues respond differently to fractionation. Fractionation allows for higher tumour dose by still low normal tissue complication rates ('biological sparing').

iv Though the image quality is inferior compared to diagnostic x-ray images because of physical reasons, the megavoltage x-ray can nevertheless be used for imaging.

art radiotherapy department. The basic connections are between CT-scanner and TPS and R&V and LINAC. The rest is 'more' and in many departments still done by film or by printout manually typing the parameters into the system next in the chain. This is time consuming and prone to errors. However, even if all modalities are connected together it is still a conglomeration of different network hardware (e.g. Ethernet, RS232), different protocols (e.g. TCP/IP, IPX) and different proprietary interfaces. The starting point of the digital therapy network was the need for an independent control of the treatment parameters and an automated record of the treatment sessions which lead to the connection between the LINAC and a separate computer with the R&V. At the beginning of the treatment planning chain, on the other hand, image information was fed into the TPS. This is where DICOM soon came into play since it is the present standard for communication with a CT-scanner. In the following steps of the planning chain many other interfaces were needed since more than merely the image information have to be forwarded. The situation differs in that from diagnostic radiology: in therapy, the image information is necessary but the machine (plan) parameters can be even more decisive. The lack of a standard led to many different de-facto standards (e.g. Impac) or proprietary interfaces. Though the history of networking may be older in therapy, the standardization is behind compared to radiology. With the publication of the final version of the so-called DICOM RT standard this may change.

3. DICOM BASED NETWORK COMMUNICATION

In the 1980's it became evident to radiologists and manufacturers of medical imaging equipment that the tremendous growth of digital image acquisition systems, archiving systems and radiology and hospital information systems (RIS and HIS) made a standard necessary for connectivity and interoperability of the different systems. This led to the development of the international standard for digital imaging and communications in medicine, DICOM[v]. It was a joint effort by the American College of Radiology (ACR) and the National Electrical Manufacturers Association (NEMA).[vi] The core of the version 3

[v] Information on DICOM can be found, e.g., on the OFFIS website[2] (University of Oldenburg) or on the NEMA website[1] itself where the original documents can be found.

[vi] The first image standard was named ACR-NEMA 2. Version 3 of the ACR-NEMA standard is called DICOM, or DICOM 3.

DICOM standard (released in 1993) is stable since a few years, but it is still extended in different directions.

3.1 DICOM RT objects

In 1999 the final versions of several new RT-specific DICOM objects were added to the standard:

- RT Structure Set
- RT Plan
- RT Dose
- RT Image
- RT Treatment Record.[1]

The first three of these new objects are directly related to the steps of the treatment planning process as pointed out in section 1.1. One may take fig. 1 as a visualization. RT Treatment Record covers the area of the R&V. Images in RT can be seen in principle as conventional radiographs. However, they are radiographs with a known projection geometry, a so-called cone geometry, as is indicated by fig. 2. One has to keep in mind that an RT image is always a visualization of the treatment portal. Therefore, it includes a central axis marker and the treatment field boundaries as well as beam modifiers such as shielding blocks, e.g. These visualizations are only valid if the projection geometry matches exactly the one of the linear accelerator.

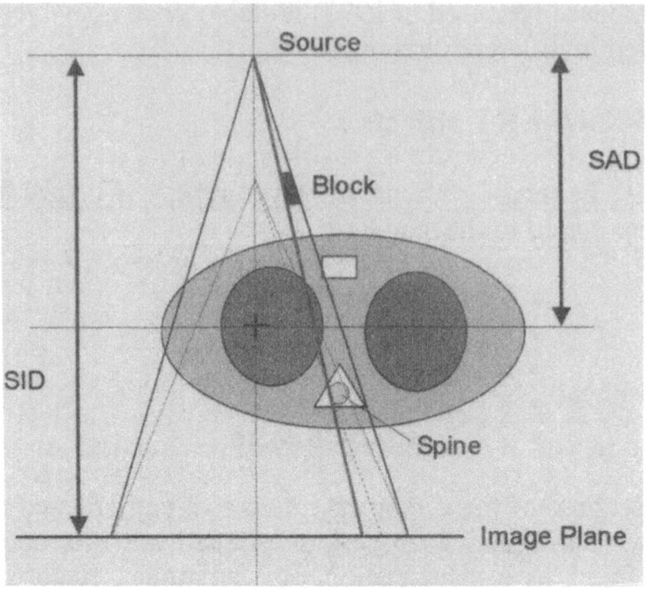

Figure 2. Scheme of the set-up of an RT Image: image with conical geometry. Shown is a situation in the thorax region. Part of the right lung is to be irradiated whereas a shielding block spares the spine. A change in the Source-to-Image (SID) and Source-to-Axis Distance (SAD) would result in a different position of the shielding block in the image plan as indicated by the dashed line.

It has to be stressed that 'DICOM RT' is not a standard of its own but part of 'DICOM' (DICOM 3). There is no principal difference between the RT objects or any other Information Object Definition (IOD) in DICOM, such as, for instance, MR image.

3.2 The State of the Art

The preliminary draft versions of IODs of RT objects were published by NEMA as early as 1996. Therefore, one could ask about the implementation of the standard in radiotherapy. The TPS may serve as an example for what can be expected from existing products at present. However, for a detailed understanding some additional DICOM vocabulary has to be introduced.

3.2.1 Service Object Pairs

The DICOM objects discussed above are worthless without knowing what to do with them, that is, with which 'service' they are connected (for example 'storage' for image transmission). They always form pairs of objects with a service (SOP). To make the situation even more complicated, the modality (e.g. TPS) speaking DICOM can act as a Service Class Provider (SCP, server) or as a Service Class User (SCU, client). Communication is only possible when a user finds a provider and vice versa. Fig. 3 tries to illustrate that.

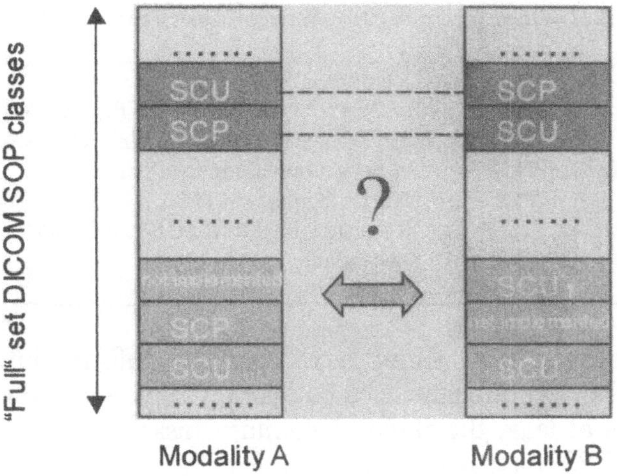

Figure 3. DICOM communication between different modalities. Communication is only possible between a user of a service class (SCU) and the provider (SCP) of the same service class (e.g. storage) in regard to the same object (dark boxes). It is indicated that "full DICOM compatibility" not always means full compatibility. (Either the SOP class is not implemented despite promises of the salesmen, or it does not provide the service 'direction' needed.)

3.2.2 DICOM Conformity of Various Treatment Planning Systems

Table 1 shows an overview of some of the most common TPS' and virtual simulation systems[vii]. The first three entries show detailed information. We learn that the Helax TMS can import structure sets (SCP) and that the Plato of Nucletron can export a radiotherapy plan (SCU) to another modality. The CMS Focus can handle both and even import RT plans (though, at the moment only from the Plato system according to fig. 3). For the Somavision, on the other hand, it is only

[vii] A virtual simulation system can be coarsely described as a treatment planning system without dose calculation.

said that it provides compatibility with RT Image, RT Structure Set and, RT Plan (brochure of the manufacturer). It is not stated whether its server or client and to which service (so in this case its mostly storage). The other two products do not provide any DICOM RT compatibility at all (status: 9/2000).[viii] Interesting enough, one vendor promises nevertheless full DICOM RT connectivity on his website.

Table 1. Information of some vendors of TPS or virtual simulation software in regard to the DICOM RT conformance of their products. (SOP: Service-Object Pair; SCP/SCU: Service Class Provider/User)

Company / Product	Type	Details (e.g. SOPs)
CMS / Focus	TPS	RT Structure Set Storage (SCP, ⇒ import), RT Plan Storage (SCP/SCU, ⇒ import/export)
Helax (Nordion) / TMS	TPS	RT Structure Set Storage (SCP)
Nucletron / Plato	TPS	RT Plan Storage (SCU)
Varian / SomaVision	VSS	"RT Image, RT Structure Set, RT Plan"
ADAC / Pinnacle	TPS	not available
Medcom / Exomio	VSS	not available

This short survey may show how carefully information of the vendors about their products have to be taken. Only one product out of our list provides at least the three elementary tasks we would like to perform with a TPS via DICOM (though, for the SomaVision it might be possible, too). It is a common experience that sales people sometimes promise features which for the programmers are merely intentions. However, it is one advantage of DICOM compared to other standards that it includes a means for checking the compatibility, the DICOM Conformance Statement.[3] Every manufacturer who claims conformity to DICOM has to provide a detailed document where he states which SOPs and with which specific implementation are part of his product. These statements are finally the basis on which the compatibility of the product with other modalities can be judged and they should be consulted before decision making.[ix]

4. TELERADIOTHERAPY

According to the publications of and the discussions within the radiotherapeutic societies[x], teleradiotherapy seems not to be an issue to

[viii] Though they do have DICOM compatibility: They can read DICOM format CT series.

[ix] Fortunately, the conformance statements of most manufacturers can be easily accessed on their websites.

[x] E.g. the Deutsche Gesellschaft für Radioonkologie (DEGRO) or the European Society for Therapeutic Radiology and Oncology (ESTRO).

date. However, one could imagine two incentives for developments in that direction. One problem in radiotherapy is a patient to be treated who has already got a treatment at another therapy unit. At least in some cases the exact extension of the previous treatment(s) is crucial for a proper and save planning of the following treatment. There could be unwanted overlaps in treatment area which would result in a higher probability of complications, especially for the so-called organs at risk. At present, the treatment plan together with images is sent as facsimile or as photocopy by mail. Whoever tried to understand a treatment record or plan documented with a different system than the one in daily use – not to speak of transmission losses – knows how difficult a task that could be. Certainly, a digital transmission using the appropriate DICOM objects could save time and facilitate the understanding of the pre-treatment and therefore ensure a save second treatment.

A second problem in radiotherapy are breakdowns and maintenance shifts of the treatment machines which are not so rare, unfortunately. Where departments are in close vicinity, it would be a great advantage just to shift the patients to a neighbouring department which has free capacity. This is of course a delicate matter among physicians in radiotherapy. There are not only technical problems but also the fear of treatment errors and legal aspects which prevented such patient transfers from coming into routine. However, there exist examples of small practices with good experiences.[4] At present, this seems only possible for departments using the same software, namely the TPS and the R&V system. With the RT extension of DICOM the technical problems at least could be overcome.

5. CONCLUSIONS

Since quite some time, radiotherapy is carried out in a networked environment. However, to date the different modalities are connected on a point-to-point basis with different standards and proprietary interfaces. The newly included RT objects will make DICOM the standard for communication in radiotherapy. At present, implementation of the RT objects for most products is still at the beginning as was demonstrated by the example of treatment planning systems. DICOM will enhance the performance of the RT network and facilitates the exchange of products of different vendors. Finally, DICOM could turn out as a prerequisite of exchange of RT information between different departments (teleradiotherapy).

REFERENCES

1. NEMA, 1999, DICOM Part 3: Information Object Definitions, pp. 58-67 a. 258-311, http://Medical.nema.org/dicom.
2. http://www.offis.uni-oldenburg.de/projekte/dicom/projekt_dicom5.htm.
3. NEMA, 1998, DICOM Part 2: Conformance, http://Medical.nema.org/dicom.
4. Dennert, P., Erfahrungsbericht: Zwei Jahre digitale Praxis und Satellit, Strahlenther. Onkol. 176 (2000) S1: 1.

The Challenge of Telepathology

SVEN SEIWERTH
Institute of Pathology, Faculty of Medicine, University of Zagreb

Key words: Telepathology, Diagnosis, Consultation, and Teaching, New developments, Croatia

Abstract: Telepathology, as a discipline of telemedicine, recently became a rapidly expanding field also in Europe. On the traces of decades of sporadic experience (first in the United States and then in some European countries) and fostered by the rapidly developing technical possibilities in communication and computers, telepathology is becoming a widely used tool. It's application seems to be of best value in countries with relatively few pathologists or with special geographical or organisational circumstances such as few reference centres or long distances between peripheral and university hospitals. Finland with its long experience in telepathology is an example of this. In Croatia the recent war and occupation of large parts of territory by Serbian troops (1991 - 95) resulting in severe communication problems, stimulated introduction of telepathology systems. Specially challenging was the need of introducing systems which could be based on old-fashioned infrastructure and used by routine pathologists. This resulted in introduction and continuos development of image database and telepathology programs ISSA and PHAROS. These systems enabled still ("frozen") image transmission with parallel voice or textual communication. In this paper the development of tools for "scanning" of still images, one of the major problems in still image telepathology is highlighted.

1. INTRODUCTION

Pathology, as a morphologic discipline offers large application possibilities for computerised tools, both in science and routine. In the past decades great progress has been made in the field of patient data archiving and retrieval, but the most intriguing advances have been made in image storage, manipulation and communication.

Telepathology, as a branch of telemedicine transfers patient data and images to remote sites for diagnostic, teaching or research purposes. It depends on many technical and human resources and is one of the most demanding telemedical fields (1).

2. GENERAL REMARKS

The diagnostic application of telepathology goes from intraoperative frozen sections, over second opinion of fellow pathologist to consultations with an expert. It is mostly used in everyday routine work in areas with few pathologists covering several hospitals, in diagnostic work with remote hospitals without pathologist or in expert consultation in difficult diagnostic cases.

As telepathology operates with images (mostly full colour) it generates lots of data, transfer of which represents one of the major technical problems.

Communication channels currently used worldvide for telepathology range from simple telephone lines to sophisticated satellite-based connections. Although communication possibilities generally improve most of developing countries will have to rely on ordinary telephone lines for years to come.

3. STILL VERSUS LIVE IMAGES

One of the major controversies in telepathology is whether still or live image transfer and analysis gives better results (2).

Live images communication gives to the receiving pathologist an opportunity of on-line scanning the material in a way similar to that on the microscope. This is especially true if the system includes a remote robotic microscope. By this way a high level of user comfort is achieved, combined with high diagnostic quality (3). To operate such a system, triple ISDN or broad band communication channels are needed, which is currently not commercially feasible in most instances. If we transmit high resolution, high quality images, only broad band channels can provide ample capacity for approximately live transmission.

Still image systems which can be operated also on ordinary telephone lines or using Internet seem to be economically feasible in many instances. One of the major drawbacks of this approach is that it can be used for transmission and analysis of limited number of high quality images coupled with a limited, time consuming, capability of scanning the specimen, actually not used in practice. This can result in serious diagnostic difficulties or even errors due to not representative sampling (2). So one of the major challenges in still image telepathology is to ensure for the receiving pathologist the possibility of reviewing the whole specimen before rendering his diagnosis.

4. INTERACTIVE VERSUS STORE AND FORWARD TELEPATHOLOGY

Still image telepathology can be performed in two ways – on-line (interactive) or by store and forward concept. Interactive communication means that the referring pathologist and the consultant are both simultaneously present an exchange notes (via separate telephone line or on a "chat" bar). Store and forward concept understands that they send/analyse the material upon convenience (certainly within reasonable limits). As most experts are professionals with tight schedules this concept seems to be more appropriate in many instances.

5. TOOLS FOR SPECIMEN SCANNING

One of the most widely used concepts is to scan the specimen in low resolution and than to focus, with high resolution, on specific areas. This approach is time consuming and to make it halfway working demands interactive communication.

We developed and tested several new tools facilitating scanning of specimens in high resolution, still image mode of transmission by applying on-line or store and forward mode of communication.

The program we use is Pharos (VAMSTECH, Zagreb, Croatia). It is a telepathology/telemedicine program working under Windows and Windows 95, Windows 98. It is designed for simultaneous transmission and storage of text and images. The communication is possible in "live" or "store and forward" mode. In "live" mode a whiteboard is used enabling multiple images to be analysed separately on consequent pages or on a single one. Interactive screen and parallel voice or "chat" communication is possible. The system can be used either on standard telephone or ISDN lines (5).

The new tools we recently tested are the "patchwork" - enabling automatic arrangement of collected images and "cine loop" enabling scrolling of the histology slide. The patchwork tool arranges automatically an unlimited number of collected images in a 3x2, 4x3, etc. manner. When the patchwork is assembled the visible size of the individual images is in reverse to their number. As each single image is transmitted in full electronic size, it can be expanded to full screen without loosing quality (5). "Cine loop" stores a continuos sequence of images from which single ones can be expanded to full size upon need by simple click. This mode is even more timesaving than the patchwork, especially in image acquisition. It brings together in an optimal way the need for specimen scanning, high resolution and store and forward mode of operation. A major drawback of "cine loop" is the need for lot of storage and communication space.

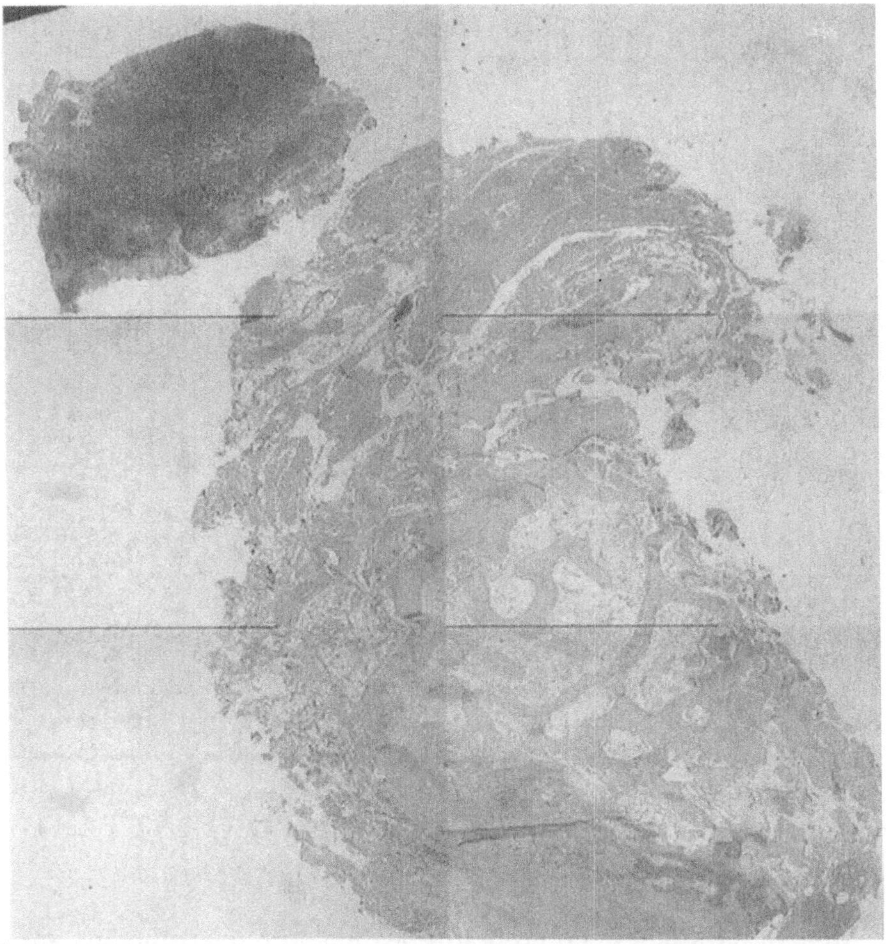

Figure 1. Patchwork image composed of six full size images enabling screening of specimen

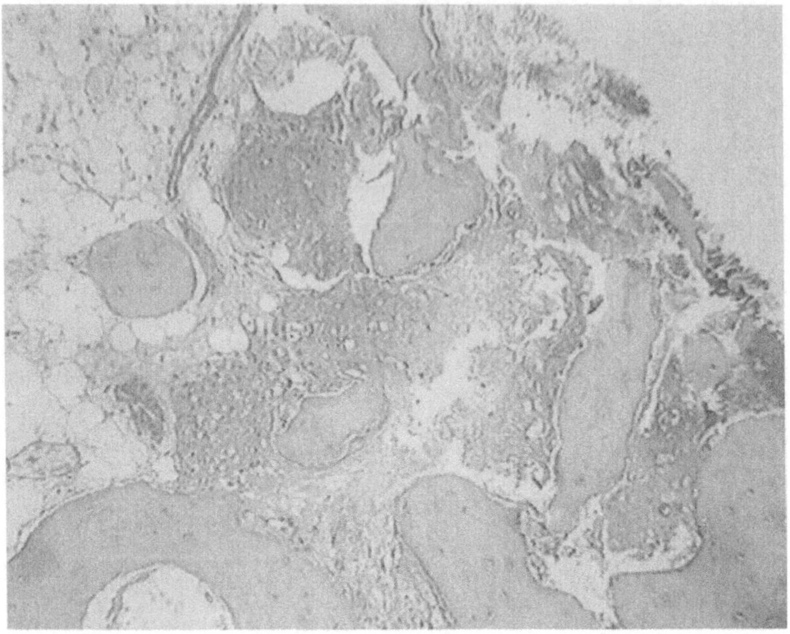

Figure 2 Full size expanded detail (marked by rectangle in previous image.

6. IMAGE DATABANK AS AN IMPORTANT PART OF TELEPATHOLOGY SYSTEM

Most of the studies and programs dealing with telepathology are so far concentrated on the question "how to transmit as many as possible representative images in the shortest time possible?" Although it is the basis for all telepathological endeavours it is by far not the only one. Issues such as communication protocol (chart), safety of patient data and documenting the final diagnosis are of great practical (legal) importance. To fulfil the goal of transmitting images together with patient data and in the same time to get a flow chart of all the communications performed, we linked the image/patient databank ISSA to the telepathology/telemedicine program PHAROS (7). In this way all patient data are stored together on the sending side, transmitted together and stored together on the receiving side. If dynamic (on-line) communication is needed, the images (together with respective data) can be placed on the PHAROS whiteboard and analysed in real time. All written communication is stored and can not be changed afterward. Also a second to second communication time chart is produced.

Data safety is relatively good because of point to point communication and can be even enhanced when personal data are sent separately linked only by code to the rest of the file. Using this approach we have the feeling that many elements important for a routinely functioning telepathology practice can be accounted for.

Teaching is one of the major goals of today's communication (5). The above-described system has also been successfully applied in the field of under and postgraduate teaching (6).

7. CONCLUSION

Telepathology represents in meanly technical demands the high end of telemedicine. In our work so far we have tried to solve some of the most crucial problems in telepathology in order to rend this discipline applicable as possible. As acquisition and transmission of images is the most time consuming part of telepathology, the store and forward concept gives both the sending and receiving side a better possibility of organising their work so that telepathology activity does not collide with other engagements. This request is, from our point of view, crucial in introducing telepathology as a routine tool. In addition, the patchwork and cine loop tools introduced by our group give the possibility of much better insight in the architecture of the specimen reducing possible misdiagnosis and fostering positive attitude toward telepathology. Linking patient/image database to telepathology system gives additional data safety and confort in managing simultaneously patient data and images.

REFERENCES

1. Weinstein R, Bhattacharzza A.K., Graham A.R., Davis J.R. Telepathology:a ten year progress report. Human Pathology 28 (1997) 1-7.

2. Dunn B.E. et al. Dynamic-robotic telepathology: Department of veterans affairs feasibility study. Human Pathology 28 (1997) 8-12.

3. Danilovic Z., Dzubur A., Seiwerth S. Concept of telepathology in Croatia. Arch Anat Cytol Pathol 43 (1995) 282-284.

4. Danilovic Z., Seiwerth S., Kayser K., Banach L., Babic D., Dzubur A. Experience based approach to interactive versus "store and forward" telepathology. Adv Clin Path 2 (1998) 149-150.

5. Barrie JM, Presti D.E. The world wide web as an instructional tool. Science 274 (1996) 371-372.

6. Seiwerth S., Jukic S., Danilovic Z., Manojlovic S., Cviko A., Banach L., Boric B. Teaching pathology using interactive image databank. Adv Clin Path 2 (1998) 157-158.

7. Dzubur A., Seiwerth S., Danilovic Z. Benefits of image databank supporting the telepathology system. Adv Clin Path 2 (1998) 158-159.

Improvement of Fetal Diagnostics Using Telepathology

K. S. WEHRSTEDT, C. TENNSTEDT, P. HUFNAGL
Department of Pathology (Rudolf-Virchow-House), Charité, Medical Faculty of the Humboldt University, Berlin

Key words: telepathology, fetalpathology, autopsy, second opinion, TPS

Abstract: In the last decades the number of described malformations and syndromes has increased rapidly and the diagnosis requires extensive experience. A main problem is the rareness of experts worldwide, so getting a second opinion is not easy. Telepathology was found to be an important method for the improvement of histological diagnosis. We will show how telepathological techniques can be used for diagnostics in fetal pathology. We analysed the telepathology system TPS within routine workflow of fetal autopsy. This system was developed in a cooperative effort by the Institute of Pathology, Charité Berlin and the Leica Microsystems Wetzlar GmbH. The main view was focused on routine capability, image quality and the communication possibilities of the system. Finally we found that the quality of fetal diagnostic can be improved by using telepathological techniques.

1. INTRODUCTION

Telepathology was found to be an important method for the improvement of histological diagnosis especially in retrieving a second opinion for difficult cases (3). The aim of this paper is to demonstrate how telepathology can be used for the improvement of fetal diagnostics.

Telemedicine: Medicine and Communication
Edited by T. M. Buzug *et al.*, Kluwer Academic/Plenum Publishers, 2001

1.1 Fetal pathology within a centre of perinatal medicine

Fetal pathology plays an important role within a centre of perinatal medicine, which is focused on healthcare of high-risk pregnancies (1). The aim of this centre is to recognise and manage congenital malformations and growth retardation, with the purpose of starting appropriate therapy at the earliest time to reduce sickness and mortality.

What are the tasks of pathology within the perinatal medicine: Firstly the pathologist is responsible for the final diagnosis of his autopsy. Secondly the quality assurance of the prenatal diagnostics lies more or less in his hands. Therefore he has to compare the reported clinical findings with the findings during the autopsy. Furthermore the cytological diagnostic of amnion fluid and fluid of body cavities as well as the diagnostic of placenta are the tasks of pathologists. An other main issue is the interdisciplinary discussion of the findings within a perinatal center.

1.2 Advanced clinical demands on fetal autopsy

In the last decades the technical and medical research in prenatal diagnostic was improved rapidly. By the use of high-resolution ultrasound and colour Doppler imaging the obstetricians are able to make more precise diagnosis and determine malformations at an earlier stage of gestation For example congenital heart defects can be detected between the 8th and 10th week of gestation. As a result of the improving quality of the prenatal diagnostic there are increasing clinical demands on the autopsy of foetuses (4). With respect to the identification of further chromosomal anomalies the molecular genetics and the cytogenetics get an increasing influence on the autopsy strategy.

1.3 Problems in fetal pathology concerning the development of the last decade

Several problems of fetal pathology result from growing clinical demands, which lead to smaller foetuses and detailed questions of the clinicians to the pathologist (5).

The current situation in fetal pathology is influenced by the following facts. In recent years the number of described malformations and syndromes has increased rapidly. Important questions often arise during the autopsy, conventional second opinion needs a lot of time, the diagnosis of fetal malformations requires extensive experience and experts of fetal pathology are rare world-wide.

If pathologist doesn't meet the increasing clinical demands, the clinical acceptation of the autopsy of foetuses will decline, particularly in specialised institutions. By the use of modern communication, digital databases and telepathology, the pathologist would be able to meet the requirements.

1.4 Telepathology offers a solution

The main application fields of telepathology today are the ways of getting a second opinion for conventional histology (mostly offline consultation) and primary diagnostic of frozen sections (online communication including remote microscope control), (6). We found a new application area for telepathology in the improvement of fetal diagnostics. The most interesting practical applications are the online consultation of an experienced pathologist during autopsy and the offline request to an expert for a second opinion in rare malformations. The second scenario is not described within this paper.

2. MATERIAL AND METHODS

2.1 The telepathology system TPS 1.5

At our department we used the telepathology system TPS (version 1.5), which was developed in a cooperation of the Charité and Leica Microsystems (Copyright Leica Microsystems Wetzlar GmbH, Charité Berlin 1998-2000) (2). This system allows direct online communication during the fetal autopsy between the laboratory for fetal pathology and the workplace of experienced pathologist. The communication protocol between TPS systems is based on TCP/IP. We realised connections via LAN, Internet and ISDN (4B-channels). The telepathology system TPS enables the investigator to save pictures taken during the autopsy in a case based database, so that they can be sent later to an expert for a second opinion via mail.

The telepathology system located in the laboratory for fetal autopsy consists of a PC (Pentium II) with a webcam, microphone and speaker, a stereomicroscope (Olympus SZH 10, Olympus Co. Japan) and a macroscope (VZ 15b, Wolfvision GmbH Austria). The telepathology system of the experienced pathologist was based on a PC (Pentium II) with a video conferencing unit. The connection between the two TPS systems was realised via LAN (10 Mbit/s) (figure 1).

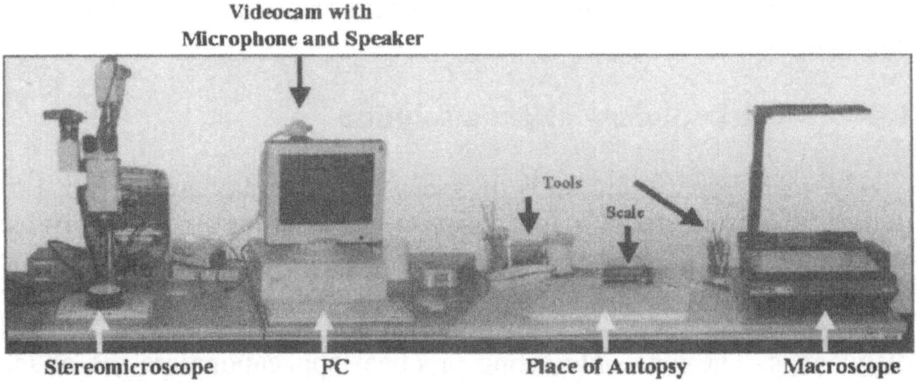

Figure 1. The laboratory of fetal autopsy with the technical equipment.

The static and the dynamic images are transferred in JPEG format. Static images have a resolution of 762 x 508 pixel (RGB PAL). Dynamic images, which have a resolution of 360 x 420 pixel, are also transferred in the JPEG format with 6 frames per second.

2.2 Routine use of the telepathology system TPS

We integrated the telepathology system TPS in the routine workflow of fetal autopsy. Within a study of 10 cases we analysed the image quality and the communication possibilities of the system. Therefore we used the telepathological consultation scenario illustrated in figure 2. The online connection to an experienced pathologist helps an unexperienced doctor to discuss and classify malformations by demonstrating his findings using macroscope and stereomicroscope. A main aim is the determination of the further autopsy strategy.

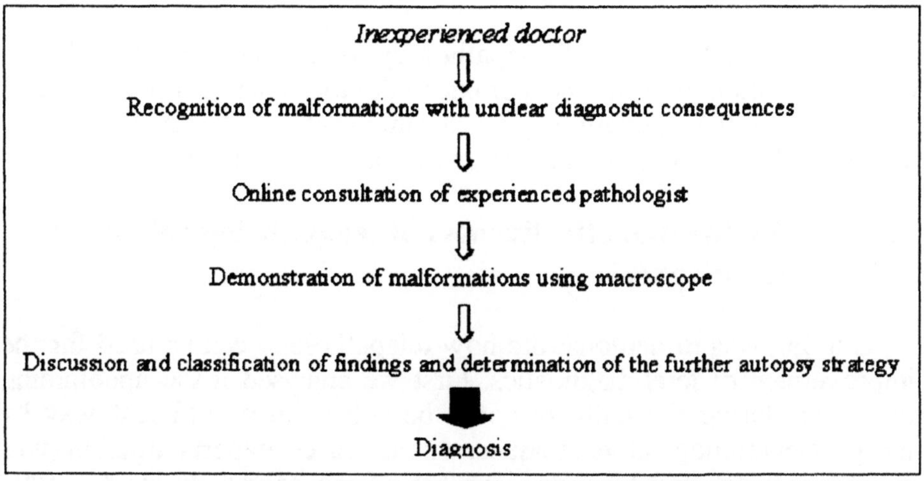

Figure 2. The telepathological consultation scenario.

2.3 Material

All cases of our study are listed in detail in table 1. We used a huge variety of cases covering a broad field of malformations, diagnoses and ages. We list the gestational age (wk) and the final diagnosis. All cases are routine cases of the department of pathology of the Charité Berlin.

Table 1. List of cases

No.	wk	Final diagnosis
1	26	Growth retardation, placental insufficiency
2	34	Trisomy 18 with VSD, abnormalities of the brain
3	23	Trisomy 21 with dysmorphic signs, VSD
4	23	Spontaneous abortion
5	23	VSD, PaVS, dilatation of PT
6	15	Trisomy 18 with horseshoe kidney
7	21	Situs inversus, HLH cystic kidney and pancreas disease
8	19	Hygroma colli, TAC type I, hypoplastic thymus
9	21	Omphalocele, dextropositio cordis, VSD, diaphragmatic hernia
10	15	Triploidy cystic kidney disease

HLH: Hypoplastic left heart; PaVS: Pulmonary valve stenosis; PT: Pulmonary trunc; TAC: Truncus arteriosus communis; VSD: Ventricular septal defect; Wk: Week of gestation

3. RESULTS

We evaluated the use of telepathology for the solution of problems during the autopsy by an unexperienced doctor. Therefore we analysed the result of the telepathological communication, the quality of the images and the time needed for the solution of the problems.

3.1 Quality and effectiveness of telepathological advice and diagnosis

Our aim was to demonstrate how telepathology can be used for the improvement of fetal diagnostics. First we analysed if the upcomming questions during the autopsy could be solved in a sufficient way by using telepathological techniques. In all cases experts opinion was really necessary and the questions have been solved during the telepathological consultation by an expert. Consequently the physical presence of an expert was not necessary. The results are listed in table 2.

Table 2. Success and time of the telepathological request

No.	wk	TP-Question	Time (min)		
			Conventional by unexp. doctor	Teleconsultation Using TPS	Conventional by expert
			estimated	measured	estimated
1	26	Dysmorphic signs, pseudocardiomegaly	70	50	20
2	34	VSD	80	70	30
3	23	Dysmorphic signs, VSD	90	75	40
4	23	Dolichocephaly	40	30	15
5	23	VSD, PaVS, PT	100	85	50
6	15	Horseshoe kidney	35	25	15
7	21	Situs inversus, HLH, cystic kidney disease	130	110	40
8	19	Hygroma colli, type of TAC, hypoplastic thymus	85	70	30
9	21	Dextropositio cordis, VSD	75	65	25
10	15	Cystic kidney disease	60	50	20

HLH: Hypoplastic left heart; PaVS: Pulmonary valve stenosis; PT: Pulmonary trunc; TAC: Truncus arteriosus communis; VSD: Ventricular septal defect; Wk: Week of gestation

3.2 Adequateness of static and dynamic imaging

The telepathology system TPS allows static and dynamic image transfer. Static images have usually a higher resolution than dynamic images. The disadvantage of static images is not having a possibility to assess motion in the scene, which is necessary for the guidance during the autopsy. The numbers of images kept by macroscope and by stereomicroscope are listed in table 3. We realised that the quality of the images was sufficient enough to make a diagnosis. In some cases dynamic imaging has been required. This depends always on the complexity of the malformation and the malformation pattern.

Table 3. Static images

No.	wk	Final diagnosis	Number of images	Images kept by macroscope	Images kept by stereo-microscope	Quality sufficient	Dynamic imaging needed
1	26	Growth retardation, placental insufficiency	16	13	3	Yes	No
2	34	Trisomy 18 with VSD, abnormalities of the brain	13	10	3	Yes	No
3	23	Trisomy 21 with dysmorphic signs, VSD	33	28	5	Yes	Yes
4	23	Spontaneous abortion	6	6	0	Yes	Yes
5	23	VSD, PaVS, dilatation of PT	28	14	14	Yes	Yes
6	15	Trisomy 18 with horseshoe kidney	9	5	4	Yes	No
7	21	Situs inversus, HLH, cystic kidney and pancreas disease	34	9	25	Yes	Yes
8	19	Hygroma colli, TAC type I, hypoplastic thymus	15	4	11	Yes	Yes
9	21	Omphalocele, dextropositio cordis, VSD, diaphragmatic hernia	13	8	5	Yes	Yes
10	15	Triploidy with cystic kidney disease	20	0	20	Yes	Yes

HLH: Hypoplastic left heart; PaVS: Pulmonary valve stenosis; PT: Pulmonary trunc; TAC: Truncus arteriosus communis; VSD: Ventricular septal defect; Wk: Week of gestation

Table 4. Dynamic imaging

No.	wk	Final diagnosis	Time of Videoconference (min)	Resolution sufficient	Image quality sufficient	Problems
1	26	Growth retardation, placental insufficiency	Not necessary			
2	34	Trisomy 18 with VSD, abnormalities of the brain	Not necessary			
3	23	Trisomy 21 with dysmorphic signs, VSD	20	Yes	Yes	No
4	23	Spontaneous abortion	5	Yes	Yes	No
5	23	VSD, PaVS, dilatation of PT	50	Yes	Yes	No
6	15	Trisomy 18 with horseshoe kidney	Not necessary			
7	21	Situs inversus, HLH, cystic kidney and pancreas disease	80	Yes	Yes	High-lights
8	19	Hygroma colli, TAC type I, hypoplastic thymus	20	Yes	Yes	No
9	21	Omphalocele, dextropositio cordis, VSD, diaphragmatic hernia	50	Yes	Yes	No
10	15	Triploidy with cystic kidney disease	10	Yes	Yes	High-lights

HLH: Hypoplastic left heart; PaVS: Pulmonary valve stenosis; PT: Pulmonary trunc; TAC: Truncus arteriosus communis; VSD: Ventricular septal defect; Wk: Week of gestation

3.3 General problems

During our study some general problems have been detected:

- It would be very helpful, if macroscope and stereomicroscope would allow remote control. This would save time.
- It is sometimes difficult to assess the real size of organs or findings because the actual zoom of the macroscope or stereomicroscope is not accessible by the TPS. (There is not always the possibility to see a centimetre grid on the background.)

Tactile information is not transferable with the telepathology system. In contrast to our expectations this was not a real restriction within our study.

The presence of an assistant during the autopsy is not necessary but it saves time and makes the presentation of findings much easier.

4. DISCUSSION

It was possible to integrate telepathology into the routine workflow of fetal autopsies. We realised that problems during the autopsy can be solved without physical presence of an experienced pathologist just by using the tools of the telepathology system TPS. We found that an autopsy assistant easies the telepathological consultation a lot. Static and dynamic imaging are both necessary for a successful telepathological consultation during an autopsy.

Telepathology opens the possibility to contact experienced pathologists for advice or a second opinion independent of their geographical distance to the place where the autopsy was performed if the accessible bandwith allows this.

This is a way to overcome the problems in fetal pathology which are a result of the developments of the recent years (rapid increasing number of syndromes, increasing demands on clinicians, small number of experienced fetal pathologists).

Consequently the quality of fetal diagnostics can be improved using telepathology. This corresponds to the results of the introduction of telepathology in other fields of pathology (2, 3, 6).

5. CONCLUSION

Telepathology is a useful tool to get advice during a fetal autopsy. It opens the possibility of a direct consultation between a less experienced doctor and an experienced pathologist. A retrieval of a second opinion would be possible in the same manner. Also professional training during interactive consultation is possible.

ACKNOWLEDGMENTS

We thank Peter Oberbarnscheidt for his technical assistance during the study.

REFERENCES

1. Bollmann, R., Chaoui, R., Kalache, K., 1997, Pränatale Diagnostik und Therapie an der Universitäts-Frauenklinik der Charité. Humboldt-Spektrum 2: 12-18.

2. Hufnagl, P, Nguyen-Dobinsky T.-N., Stock, B., Fiedler F., Dietel M., 2000, Das Telepathologiesystem TPS 1 für Routine-Telepathologie. Pathol Res Pract 2000; 196: 450.

3. Kayser, K., Szymas, J., Weinstein R., 1999, Telepathology, Springer Heidelberg, New York.

4. Tennstedt C, Hufnagl P, Nguyen-Dobinsky TN, Chaoui R, Bollmann R, Urban M, Körner H, Tietz S, Tietz P, Dietel M., 1997, Moderne, an klinischen Anforderungen orientierte Fetalautopsie unter Nutzung von informationstechnischen Methoden, Pathologe 6: 445-452.

5. Tennstedt C, Wehrstedt K, Dietel M, Vogel M, Hufnagl P., 2000, Use of telepathology in fetal diagnostics, Pathol Res Pract 2000; 196: 442.

6. Wells, C. A., Sowter, C., 2000, Telepathology: a diagnostic tool for the millenium? J. Pathol 2000 191, 1-7

CLINICAL APPLICATIONS

Telearchiving for Craniosynostosis Research

[1]M. REICHERTS, T. H. JÜNGER, T. POSSELT, H.-P. HOWALDT AND [2]A. KRIETE

[1]*Department of Maxillo-Facial Surgery, Justus-Liebig-University, Medical School, Klinikstraße 29, D-35385 Giessen, Germany*

[2]*Image Processing Laboratory, Institute of Anatomy and Cell Biology, Justus-Liebig-University, Aulweg 123, 35385 Giessen, Germany*

Key words: craniosynostosis, telearchiving, surgeons network, standardized documentation, multi-center study

Abstract: This telearchive is used to establish a network for craniofacial malformations based on a standardized documentation of clinical findings of patients with craniosynostosis. A database system is required where standardized findings and images can be stored. For cooperation and information exchange it is necessary that all participants have access to the database. This requires a networkable system which can be handled easily for archiving and further treatment of the visual findings. We use the Client-Server-based database system Cinebase 1.5 for storage and management of these digital data. The Server is located on a SGI platform and there are SGI or Macintosh Clients available.The digital patient record consists of photos, diagnostic pictures, data from the documentation form, different anthropometric measurements and results from genetic investigations. More than 100 cases with over 300 images are archived within the cooperation of ten national university hospitals.

1. INTRODUCTION

Premature fusion of calvarial sutures, craniosynostosis, can occur isolated or in combination with other symptoms described as clinical defined syndromes. The incidence of these diseases is low (1:5000 on average) and a multidisciplinary approach to diagnosis and therapy is needed. While certain surgical procedures in craniosynostoses are well established, a standardized homogeneous data collection and documentation is difficult in the absence of common reporting standards.

Supported by recent advances in medical genetics which detected parts of the molecular pathology of craniosynostosis, standards for documentation were established by a group of national craniofacial centers. To facilitate an exchange of experiences and increase the number of patients a database for storage of multimodal clinical data was established. Based on a Client/Server-system the Telearchive Craniosynostosis enables an exchange of discussion and information for the participating hospitals.

2. DOCUMENTATION SYSTEM REQUIREMENTS

The documentation of clinical observations and findings is done with the help of a categorized classification, where frequent findings are reflected by different categories in a reporting form. Further findings are added in a plain text description. The documentation form includes pretherapeutical findings, therapeutical data, anthropometric measurements and results of genetic examinations. Beside these categorized and numerical data there are X-rays, CT's and clinical photographs stored as well. Blood samples are taken from each patient for genetic investigation. In cases where mutations are detected, a genotype-phenotype correlation is performed.

The cooperation of several hospitals implies the use of a database system where categorized findings, numerical findings and images can be stored. To obtain a considerable degree of cooperation and information exchange it is necessary that all participants have access to the database. This requires a networkable system which can be handled easily for archiving and further treatment of the visual findings. An easy selection of the desired images and fast access to the data is required. Since a majority of the data consists of images, this project can only be realized by using a database system which is able to manage a large number of pictures linked to key words, plain text or numerical values from the documentation form.

3. THE TELEARCHIVE CRANIOSYNOSTOSIS

Based on the documentation system for craniosynostosis we have introduced the Telearchive Craniosynostosis, which contains multimedia files of these patients. To realize this telearchive, we decided to use the Client-Server-based database system Cinebase 1.5 (Cinebase Software, L.A.) for storage and management of these digital data. The server is able to manage different digital data (pictures, videos, 3D models), and the media can be stored in many different formats (TIFF, GIF, JPG, PDF). Within this system the user can create metadata, which establishes the relationships between media files and descriptional data. By a user defined index with multiple references it is possible to associate text, numerical information, keywords or hierarchies to media files. The index is based on the documentation form mentioned above. Hierarchies are created by using available data like skull shape or other findings. In this way, it is easy to retrieve assets from the database through a wide variety of query services quickly and intuitively. The digital patient record consists of photos, diagnostic pictures, data from the documentation form, different anthropometric measurements and results from genetic investigations. Fig. 1 shows a typical screen shot of the data for one patient, including the documentation data and previews of the available media data.

The database is a Client/Server based system, the Server is located on a SGI platform. The hardware makes it possible to retrieve pictures quickly and with high resolution, even if they occupy several megabytes of storage space. There are both SGI or Macintosh Clients available. By using any TCP/IP compatible communication network, a connection to the SGI Server can be established (fig. 2). Data security is mainly achieved by the use of a properitory client software. Permissions are given by the administrator for data access of each user.

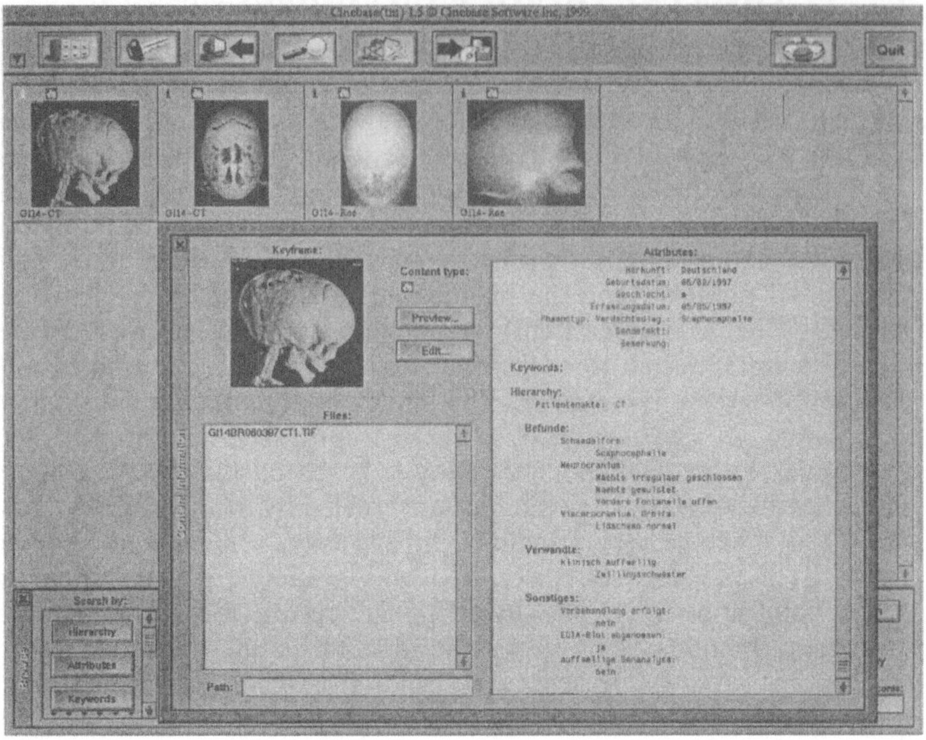

Fig. 1: Screen shot of a multimedia patient file in the telearchive craniosynostosis. Previewing of images and viewing corresponding patient data can be done simultaneously.

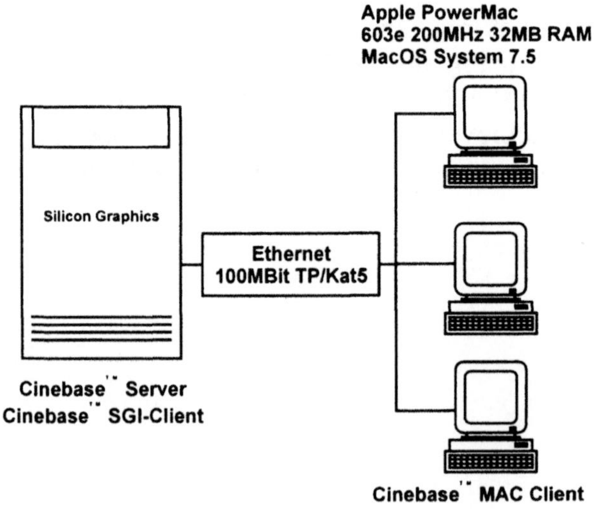

Fig. 2: Client-Server-system.

4. CONCLUSION

Based on a standardized documentation of clinical findings of craniosynostosis patients the described telearchive is used to establish a network for these craniofacial malformations. Thus, with the help of genetic investigations a genotyp-phenotyp correlation may introduce a molecular classification in the future. The telearchive is located at the Justus-Liebig-University Giessen, Germany. Since the beginning of the project in 1999 more than 100 cases with over 300 images are archived within the cooperation of ten national university hospitals. At an expected number of approximately 100 patients per year, sufficient data will be available soon to carry out studies which have the aim to find statistical correlations within the data.

ACKNOWLEDGMENTS

The work was supported by grants from the Deutsche Forschungsgemeinschaft (HO 1206/5-1).

REFERENCES

1. J.B. Mulliken, M.L. Warman (1996), Molecular genetics and craniofacial surgery, Plast Reconstr Surg, Vol. 97, No. 3, pp666-675
2. D. Steinberger, J.B. Mulliken, U. Müller (1996), Crouzon syndrome: previously unrecognized deletion, duplication and point mutation within FGFR2 gene, Hum Mutat, Vol. 8, No. 4, pp386-390
3. U. Müller, D. Steinberger, S. Kunze (1997), Molecular genetics of craniosynostotic syndromes, Graefes Arch Clin Exp Ophthamol, Vol. 235, No. 9, pp545-550
4. D. Steinberger, G. Vriend, J.B. Mulliken, U. Müller (1998), The mutations in FGFR2-associated craniosynostoses are clustered in five structural elements of immunoglobulin-like domain III of the receptor, Hum Genet, Vol. 102, No. 2, pp145-150
5. D. Steinberger, T.H. Jünger, H.-P. Howaldt, P. Christophis, U. Müller (1999), Mutation of FGFR2 (cys278phe) in craniolacunia and pansynostosis, J Med Gen, Vol. 36, No. 6, pp499-500

6. T.H. Jünger, M. Reicherts, P. Christophis, D. Steinberger, H. Collmann (1999), Klassifikation von Kraniosynostosen durch klinische Dokumentation und Zuordnung zu Genmutationen, Posterpreis, 49. Jahrestagung der DGMKG, Basel, Juni 1999

7. J.B. Mulliken, D. Steinberger, S. Kunze, U. Müller (1999), Molecular diagnosis of bilateral coronal synostosis, Plast Reconstr Surg Vol. 104, No.6, pp. 1903-1615

8. T.H. Jünger, D. Steinberger, P. Christophis (2000), FGFR-assoziiertes Cranio-synostose-syndrom mit Craniolacunia, Schädelbasiskongress, Al-Budoor Verlag, Damaskus/Syrien, pp.148-151

9. T.H. Jünger, M. Reicherts, D. Steinberger, H. Collmann, B. Kotrikova, J. Zöller, H.-P. Howaldt (2000), Standardized evaluation and documentation of findings in patients with craniosynostosis, Journal of Cranio-Maxillofacial Surgery, in press

A Case Study for Digital Systems Integration and Telemedicine in the Management of Lung Cancer Treatment

Digital Technology & Telemedical Applications

[1]N.SHASTRY, [2]A.MORGAN, [3]P.RICHARDSON and [1]M.R.REES

[1]*Department of Radiology-Division of Cardiac, Anaesthetic & Radiologic Sciences, University of Bristol & the Bristol Royal Infirmary- UBHT NHS Trust, Bristol, UK*

email: n.shastry@bris.ac.uk , Fax: 44 (0)117 9282319, [2]Department of Cardio-thoracic Surgery, Bristol Royal Infirmary, UBHT NHS Trust [3]*Department of Clinical Radiology, Bristol Royal Infirmary, UBHT NHS Trust*

Key words: telemedicine, digital radiography, ethernet, x-ray, film digitiser, DICOM, PACS, lung cancer

Abstract: It is imperative that cancer care management protocols are established rapidly to help assess, prepare and instigate the required treatment procedures. We have addressed one such requirement i.e. in the management of lung cancer care using digital systems integration and telemedicine. With the telemedical system in place 160 patients have been imaged by the Digital Radiography (DR) system and the images have been transferred simultaneously on to the viewing systems for radiology and surgical opinions. The images acquired and transmitted digitally comprise a patient group where around 80% of the patients have had cancer, needing surgical intervention; the images were both pre and post operative images.

1. INTRODUCTION

Digital imaging systems lend themselves to telemedicine whereby the requirement to digitise hardcopy films or analogue video is obviated. Systems integration solutions used judiciously would allow for quicker treatment to the patient while also bringing the treatment management costs down. Telemedicine provides an ideal method of integrating treatment management protocols at one expert centre.

1.1 Initial Comments

The existing facility for lung cancer care was based on the medical staff moving around different hospitals to examine the patients while awaiting the radiology reports, by post. To summarise the procedure: *"the general practioner (GP) or the physician would instigate the X-ray---a radiologist would then send the report back to where the request came from---the GP or the physician would consult with the surgeon --- an appointment would then be sent to the patient---the patient would go either to the physician or the surgeon as necessary".* This complete process would take between 15-30 days at least. The patient would get his response from the GP within the 15 days stipulated by the UK Government for Lung Cancer Care, while waiting for the specialist appointment. With the two simultaneous lines of digital communication established from the DR system to the radiologist and the surgeon, the response time has been reduced substantially. The surgeon, on viewing the digital image, responds straight away and books the patient in for the required procedure while waiting for the radiologist's report; there by establishing the protocol for rapid lung cancer care by reducing the wait and anxiety for the patient for further management.

2. CURRENT AND IMPLEMENTED SOLUTION

The patient sees the General Practioner (the first point of contact for the patient, here in the UK) or visits the Accident & Emergency Department for the initial urgent attention as the case may be. Both these points of contact invariably request an X-ray. The GP may also refer the patient directly to a consultant physician or a consultant surgeon. However, what transpires is that the patient has to have an X-ray. This X-ray is then reported by the radiologist and *only the report* is sent on to the referring consultant physician or the consultant surgeon. The consultant physician may also send the patient over to the surgeon if necessary. The time taken for these procedures is about 3 to 4 weeks at least while the patient is waiting for the management protocol to be implemented and for the next communication from the hospital or the GP.

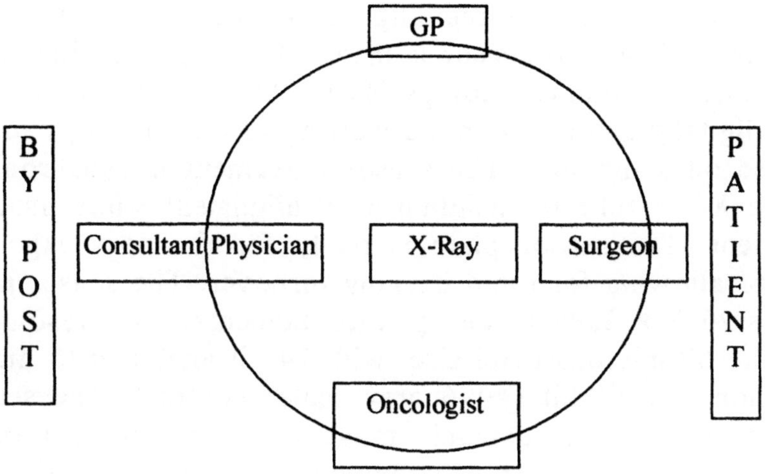

Figure 1: Image illustrates the existing, initial patient & patient data pathway and points of contact for Medical Care

The consultant surgeon has 10 remote clinics spanning a radius of 50 miles, which he visits at least once a month, spending a lot of time travelling. On an average around 74 patients per month from these 10 clinics are seen and a third of these may be sent to the base hospital, the Bristol Royal Infirmary (BRI) for a surgical procedure whereas on some occasions, the patient has to be sent back without a surgical procedure being necessary or possible.

To overcome the associated difficulties where the expert and the patients have to travel for providing and receiving medical attention we established a telemedical facility using the inherent potential of a digital imaging device the Digital Radiography system. Digital Radiography systems have been in use and have been assessed for their performance capabilities in the imaging of bone[1] and chest and lung diseases[2].

2.1 Description of equipment and the telemedical facility

The CDXI-11 X Canon Digital Radiography (CDR), system is a small footprint system and weighs less than 250kg. The CDR can work with conventional 230V/240V power circuits allowing for ease of implementation. The CDR has a LANMIT♦ digital sensor and an X-ray source aligned to the sensor. The 17" (43cm) by 17" (43cm) sensor is made of amorphous silicon (a-Si) and uses thin film technology (TFT). The sensor captures the X-rays digitally and allows rapid viewing of the soft copy image displayed on a monitor. The sensor movement is synchronised with the X-ray tube to maintain axial alignment while imaging the patient. The sensor plate is ready for the next exposure within 6s allowing for rapid imaging turnover. The sensor plate captures 2688 x 2688 pixels (picture elements); at a resolution of 160 x 160 microns pixel size, with 14 bit analogue to digital conversion and 12 bit (4096 grey scale) contrast. The sensor plate covers a larger dynamic range than conventional films. Figure 2 shows the various component elements of the DR system.

♦ Large Area New MIS & TFT sensor

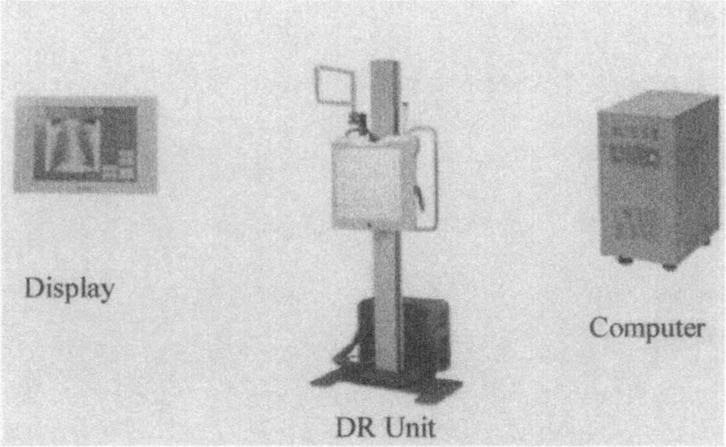

Figure 2: Components of the DR System

The control unit is a Windows NT® based dual processor computer with standard personal computer accessories such as a keyboard, high quality display monitor, hard disk, CD Rom, floppy disc and an Ethernet card for digital image data transfer. The digital data so obtained can also be piped to other PC-based viewing systems for radiology reporting, viewing systems in clinics, operating theatres and other hospitals for an integrated patient management protocol.

The computer has an Ethernet 10/100 baseT data transfer capability, which means that the digital data information can be transferred in less than 30s at 10-100 mega bits per second over intra and internet connections to hard copy output devices such as DICOM laser imagers, remote PC based viewing stations and or PACS (Picture Archiving & Communication Systems). The images are DICOM 3.0 images allowing for compatibility of viewing systems across different medical establishments in a standardised format. The hospital LAN & Wan provide the conduits for the image transfers. Figure 3 shows the architecture of the DR system and the viewing system. Figure 4 shows the existing, implemented facility and the scope for future modifications and implementation of the remote hard copy digitising facilities where the DR systems are not yet implemented.

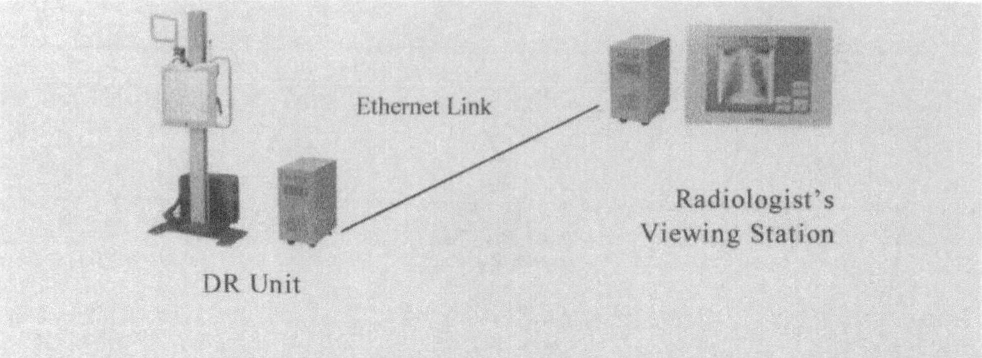

Figure 3: Architecture of DR Systems and Viewing Stations

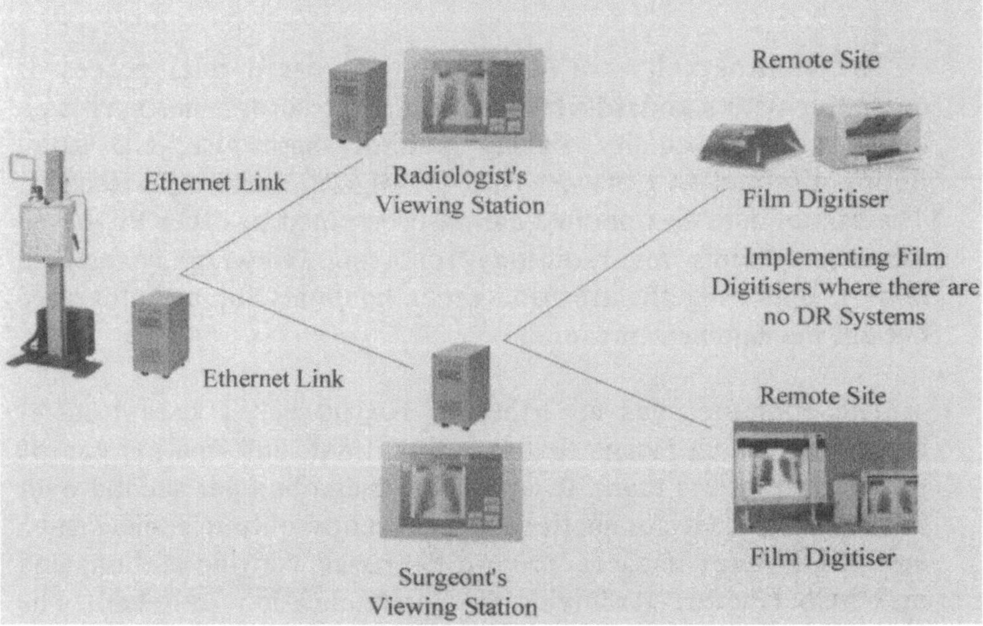

Figure 4: Existing telemedical link between the DR System and the Surgeon's PC and the future implementation of Film Digitisers and links where DR Systems are not yet available

The innovative facility in use here at Bristol serves both pre and post interventional management needs for the surgeon. A conventional chest X-ray system, Philips Medio 65 CP-H was used as the source while the digital radiography system replaced the conventional film.

The chest images from patients were "captured" in a digital format, DICOM3. This enabled the digital image data to be transmitted simultaneously to the radiology viewing system for the radiologist's report while also being transmitted to the interventionalist, the cardio-thoracic surgeon in this case. While awaiting the report from the radiologist, the surgeon would immediately instigate the admission protocols for a surgical procedure, as necessary. This reduced the normal 3-4 weeks waiting time for the patient, enhancing the early implementation of the patient management protocol while also alleviating the patient's distress and anxiety.

3. FUTURE POSSIBILITIES AND SCOPE FOR RESEARCH

The digital image data acquired has enough information for the surgeon or the end user to evaluate and analyse regions of interest in greater depth. Using appropriate security measures for maintaining patient and data confidentiality other digital communication techniques can be evaluated. Transmission speeds of digital image and text data can be improved nearly ten fold by the implementation of DSL Digital Subscriber Line communication data pathways. Remote hospitals, where DSL is not yet available the implementation of this telemedical link can be through ISDN2, bearing in mind the limited scope transfer speeds. DSL communication links and other possible large data bandwidth system implementations can be researched further to have active video conferencing as well as data transfers. This would allow for further enhanced interaction between the Lung Cancer Care Management Team. Implementation of portable digital data access facilities using portable computers can be assessed. Web or Internet based solutions can be explored further implementing ***e***-strategy in healthcare.

4. CONCLUSION

The digital image transfer system (between the DR unit, the radiology workstation and the surgeon's viewing system) for integrated patient management has been in use for 6 months. We wish to develop this concept further and extend it to the other nine referral centres for this surgeon. The same image transfer model can be extended to all other consultants here at the BRI within the Cardio-Thoracic Surgical Directorate and their associated peripheral clinics. This is the tip of an iceberg; the same image transfer model can be applied inexpensively in most medical and surgical specialities throughout the National Health Service (NHS).

ACKNOWLEDGEMENTS

Dr N.Shastry acknowledges the support given by Professor T.M.Buzug, Philips Medical Systems UK, Kodak and the organisers of Medicom 2000 for extending their kind invitation.

REFERENCES

1. Volk M, Strotzer M, Holzknecht N, Manke C, Lenhart M, Gmeinwieser J, Link J, Reiser M, Feuerbach S, 2000. *Digital radiography of the skeleton using a large-area detector based on amorphous silicon technology: image quality and potential for dose reduction in comparison with screen-film radiography.* Clinical Radiology, 55,615-621, 2000

2. Bernhardt TM, Otto D, Reichel G, Ludwig K, Seifert S, Kropf S, Rapp-Bernhardt U 1999. *Detection of simulated interstitial lung disease and catheters with selenium, storage phosphor and film based radiography*, Radiology, 213, 445-454, 1999

Neurovisual Rehabilitation via the Internet

Teletreatment and Teleproctoring of a Home Self-adaptive Visual Field Treatment on PC/ TV Screen

[1]FRITZ SCHMIELAU, [2]EDWARD K. WONG JR AND [1]FABIAN HOLBE

[1]Institute for Medical Psychology, Medical University of Lubeck, University Clinic Lubeck, Ratzeburger Allee 160, 23538 Lubeck, Germany, email: schmiela@medinf.mu-luebeck.de, Fax: 49 (451) 500-6097 [2]Department of Ophthalmology, University of California Irvine, Medical Plaza, One Medical Plaza Drive, Irvine, CA 92717, USA, email: ekwong@uci.edu, Fax: 1 (949) 824-4015

Key words: neurovisual rehabilitation, at-home treatment, hemianopia, teletreatment, teleproctoring, visual plasticity, restitution of visual functions

Abstract: Visual disorders due to cerebrovascular accidents or traumatic brain injury represent a severe individual and economic problem causing considerable constrains of personal autonomy and binding of public money. In contrast there is little treatment available which allows for true restitution of visual functions. Since within recent years the efficacy of a self-adaptive neurovisual rehabilitation algorithm had been demonstrated for an automatized treatment on a special perimeter hemisphere (LRP), the algorithm was adapted for stimulus presentation on PC and TV screens. Due to fast increasing internet access in many countries, a study has been initiated to use our "Vision Trainer Brightness" VTB software for at-home neurovisual rehabilitation teletreatment via the internet. The advantages of that approach are demonstrated in terms of time, actuality, individuality, comfort, broad treatment access and low costs. Teleproctoring of at-home treatment may serve as an additional tool to further improve treatment quality.

1. INTRODUCTION

It has been estimated by the National Advisory Eye Council (1993) that there are more than three million people in the US who are visually disabled (low vision or blind), blind: 200,000. Leading causes of low vision and blindness are diseases that are common in old age: maculopathy, cataract (400,000), glaucoma (242,000), diabetic retinopathy and optic nerve atrophy. Two to three percent of the general population suffers from amblyopia („lazy eye“).

1.1 Visual Disorders Caused by Cerebral Insults

In addition there exists a large number of people which suffer from visual disorders due to cerebral insults, such as stroke, haemorraghe, trauma and tumors. With the exception of aphasic and motor disorders, homonymous visual field defects constitute one of the most frequent consequences of cerebral insults[1]. After posterior infarctation 33% demonstrate a homonymous hemianopia, after closed brain trauma 37% and after intracerebral tumors 22%.

1.2 Stroke and hemianopia

Stroke patients make up the largest group among hemianopic patients: Out of 258 patients with homonymous visual field defects 65% had suffered infarctation of the posterior and/or medial cerebral artery, 12 % brain trauma, 9 % haemorraghe , 6 % operated occipital tumors and 4% chronic cerebral hypoxia[2].

The prevalence of stroke survivors in the US today is about 4 millions, the incidence of first-ever strokes is 500,000. 1 million people over the age of 15 suffer from disabilities resulting from a stroke. The lifetime cost of stroke is $ 40 billion for first-ever stroke only and not including indirect costs related to caregiver loss of earnings and depression[3]. The world-wide incidence rate of stroke ranges from 1.5 to 3.0 per 1,000 people (USA 2.4[4] Germany 2.0)[5]. Out of all patients in Germany who had been working before their stroke, 29-36% returned to their work, 31-52% were handicapped and suffered from a reduced autonomy and 12-21% were in-patients in clinics or homes for a long period[6]. Of all patients with cerebrovascular infarctation requiring therapy in a rehabilitation center about 20 – 30 % have homonymous visual field defects[7]

1.3 Visual Disorders and Costs

These data may demonstrate that 1. visual disorders of all kinds are an essential problem in all countries 2. brain-related visual disorders after stroke, especially homonymous hemianopia is a severe problem of the working population causing considerable temporary or permanent constrains of autonomy in most of them 3. Loss of working power and health-care costs of in-patients after primary treatment in stroke units or neurological clinics represent an economic burden to insurances and society.

2. NEUROVISUAL REHABILITATION

Chances of recovery of visual functions after cerebral lesions are discussed controversely[8 9 10]. In spite of numerous reports [11 12 13 14] and every day clinical experience on *spontaneous* improvements of disturbed visual functions, such as acuity, form and color vision and visual field defects, within some time after cerebral damage, some authors argue that visual cortex lesions result in permanent damage[15 16]

Authors supporting the view of functional recovery, limit spontaneous improvements to a period of two years after the incident[17]. Among them there exists agreement that most dramatic effects occur during the first three to six months. From the early work of Poppelreuter[18] it is emphazised that improvements of visual functions beyond that interval may only be accomplished by treatment.

2.1 Developing Neurovisual Rehabilitation of Hemianopia: From Manual Stimulus Presentation to Automatized Treatment

It was only in 1978[19 20] that Poeppel and coworkers were able to demonstrate the efficacy of treatment on the size of visual fields (increase of several degrees) in hemianopic patients with lesions of the post-chiasmatic visual system. They were repetitively presenting threshold luminance stimuli in the transition zone between the intact and anopic part of the visual field, using a Goldman-type manual perimeter. In 1979 Zihl and von Cramon[21] claimed that an average of 12.5 sessions of repetitive threshold stimulation with white stimuli were sufficient to even restitute central scotoma and induce recovery of a variety of visual functions.

2.1.1 Semi-automatized Treatment

In 1990 Schmielau and Potthoff[22] replicated the findings of Poeppel, Zihl and von Cramon in principle. Instead of 12.5 treatment sessions, an average of 50 sessions was found to be appropriate to partially restitute visual fields in hemianopic patients. In contrast to the euphoric results of Zihl and von Cramon (1979) restitution appeared to be restricted to recovery of visual fields of *white* light. Improvements of different visual functions (acuity, color vision, form discrimination) were seen only during intervals of *spontaneous* recovery within the first six months after cerebral damage or as a result of a *specific* treatment.

Moreover, semi-automatic sensori-motor treatment turned out to be more efficient and patient-comfortable in treating hemianopia than passive manual threshold stimulus presentation[23]:
Advantages of the super-threshold sensori-motor technique (including breaks every 5 min) were

- a high level of attention throughout treatment sessions
- a very low drop-out rate of out-patients during the 6 months' treatment
- therapy was practicable by technical assistants instead of scientists or doctors.

The remaining disadvantage of this procedure was the *manual* displacement of light emitting diodes (LED) serving as light sources. LEDs were attached to a perimeter arc (r = 1 m) which itself had to be rotated maually around an axis through the patient‘s fixation point, to reach any location within the binocular 180^0 visual field.

2.1.2 Automatized Self-adaptive Neurovisual Treatment Using an LED-based Prototype of a Visual Treatment Perimeter

A completely automatized treatment of neurovisual disorders was possible, when in 1990 a first *scientific* prototype of an LED-based reaction perimeter was invented by Schmielau and constructed by Wiegman[24]. Legal protection of registered design[25] and German[26], European[27] and US[28] patents were granted for that invention[29]. In 1996 a first *commercial* prototype, the Lubeck Reaction Perimeter (LRP), was constructed[30]. LRP software[31] contains a *self-adaptive* treatment algorithm, restricting repetitive stimulus presentation to the *actual* transition zone between the intact and anopic visual field within any treatment session. Several studies demonstrate treatment efficacy when the LRP was used:

In a pilot investigation[32], 6 out of 11 patients suffering from visual disorders caused by cerebrovascular or traumatic brain injuries, demonstrated visual field enlargements of more than 20^0 after 50 treatment sessions.

Within a longitudinal study[33], successful partial visual field restitution was obtained in 28 of 30 stroke patients(average age: 54 yrs) with homonymous hemianopia.

In another treatment follow-up study[34] on 80 patients (age: 17-81 yrs) with sustained post-geniculate damage caused by a variety of sources, significant and stable improvements of visual functions were demonstrated in the majority of patients. It was shown that most of the treatment effects were successfully transferred into visually guided activities of daily living (ADL).

2.1.3 TV- and PC-screen Based Treatment

Within his PhD thesis, at the Lubeck Institute of Medical Psychology, Kasten developed diagnostic and treatment software[35] for Commodore C 64 computers and treated 11 hemianopic patients presenting stimuli on a TV screen in their homes[36]. He was able to demonstrate significant improvements of brightness, form and color perception in 9 of 11 (brightness) and 8 of 8 (form, color) patients after 30 – 200 one hour treatments. This software has been adapted to the new generation of > 286 PCs and in modified form for Pentium > I PCs. In another study[37] successful treatment of 38 patients with lesions of the pre- and post-chiasmatic visual system was shown with that software (NovaVision VRT)[38]: In contrast to *large* visual field improvements seen with a diagnostic program on a PC screen (High Resolution Perimetry HRP), visual field increase was *small* in the group of patients with *pre-chiasmatic* lesions ($+ 2.1^0 \pm 0.5$) and *very small* ($+ 0.4^0 \pm 0.3$) in the group of patients with *post-chiasmatic* injuries, when tested with conventional threshold strategies, using the Tubingen Automatic Perimeter TAP 2000. The very small improvements seen in post-chiasmatic patients treated with VRT software, in contrast to the large treatment effects obtained by Schmielau and coworkers[32 33 34] on similar patient groups, have been discussed in terms of different software algorithms used for stimulus presentation on VRT-controlled PC screens and in the LRP perimeter hemisphere. [39].

3. "VISION TRAINER BRIGHTNESS" (VTB) A NEW SOFTWARE FOR NEUROVISUAL REHABILITATION ON PC

To improve the efficacy of PC-based neurovisual rehabilitation and use it for teletreatment via the internet, a new software ("Vision Trainer Brightness" VTB) was developed in Lubeck, adapting the efficient features of the self-adaptive treatment algorithm used with the LED-based LRP perimeter hemisphere. "Vision Trainer Brightness" VTB is a software for treatment of visual fied defects by means of a 19" PC or large high quality (e.g. SONY) TV screens.

3.1 VTB Treatment Algorithm

The basic idea is to stimulate the transition zone between the intact and the anopic visual field by presenting an *acoustic attention pre-stimulus*, followed at random interval (e.g. 500 – 2000 ms) by a *light stimulus* of a certain form (F), size (S), luminance (L) and duration (Δ t) on the patient's screen having a specified background luminance (L_B). Besides others, an important treatment variable is *contrast* (C) which may be defined as

$$C := L - L_B / L_B \qquad \text{or} \qquad C := L - L_B \; / \; L + L_B$$

The patient's response time *(simple reaction time, SRT)* of pressing a keyboard- or mouse-key to *detection* of short (e.g. 100 ms) light stimuli, while fixating a permanent fixation spot, is measured:

If a reaction time SRT

- is **within** a pre-set SRT interval (e.g. 150 – 900 ms), the light stimulus location will proceed within the same line/row a specified number of pixels (corresponding to a certain visual angle φ[1]) in a pre-set *direction* (right, left, up, down) towards to anopic field
- is **outside** (<, >, >>) the pre-set SRT interval or in case of **no response** several times (usually n = 2), the stimulus will jump backwards a pre-set number of pixels (usually corresponding to a visual angle = 5 x φ) (fig. 1).

The procedure of moving the stimulus towards the anopic field and jumping back into the intact field (**crossing the visual field border in both directions**), is repeated several times (pre-set value: e.g. n = 3) within a line/row. Then the stimulus location is displaced (up- or downwards / to the left or right) to the starting location within the adjacent line/row.

Breaks may be inserted manually at the patient's convenience, or automatically every 3-5 min, to prevent fatigue.

A **treatment session ends** when

- the *treatment area* on the screen or the whole screen has been stimulated
- a pre-set *number of stimuli* (e.g. n = 700), has been presented.

At the end of a session the total numbers of correct and incorrect (early, prolonged and missing) reactions, as well as the mean reaction time, standard deviation and locations of any stimulus causing correct or incorrect responses are calculated and tabulated on the screen. In addition, correct and incorrect responses are depicted as a map with

[1] φ (range: 1-3^0) is approximately equal to arc tan of the quotient of the horizontal (or vertical) distance between adjacent stimulus locations on the screen and the distance **d** between the screen and the patient's eye. **d** has to be kept constant during treatment and in-between sessions. Its magnitude has to be calculated according to d = a tan α, where e.g. α = 30^0, and **a** is the visible horizontal width of the screen. **d** should always be > the lenghth of the patient's arm (≈60 cm) to avoid accomodation problems.

respect to the respective stimulus locations of the treated screen (visual field) sector.

This type of **feedback after each treatment session** has been shown to have a major influence on the patient's motivation to continue neurovisual rehabilitation treatment[40] and in part explains our low drop-out rate.

All calculated data of a single treatment session may be compared with those from previous ones and depicted and printed as „treatment progress curves". All data are available as black or color print-outs on appropriate ink jet or laser printers.

3.2 VTB Home Treatment without Internet Access

All data are stored on a program discette used by out-patients in case of home treatment without internet access. These discettes are mailed in monthly intervals to the supervising praxis / clinic / rehabilitation center, where they are evaluated by the treating doctors. After eventual modifications of treatment variables (stimulus parameters, background luminance, treatment areas, ...) to improve efficacy, discettes are re-mailed to the patient's home to proceed with treatment. This procedure is continued until stable improvements are obtained and/or re-diagnostics of the actual visual field size with perimeters has to be performed.

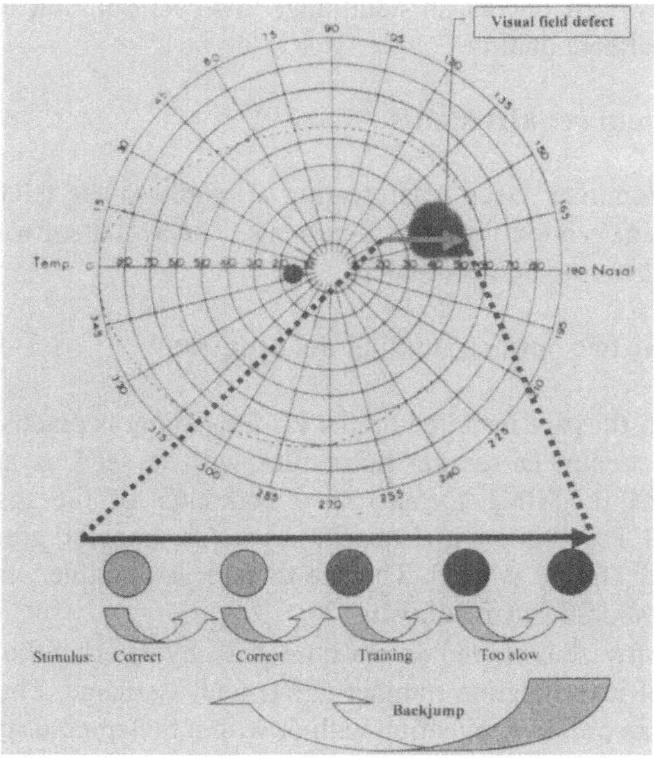

Figure 1. Algorithm of neurovisual rehabilitation treatment used with "Vision Trainer Brightness" VTB Software and the "Lubeck Reaction Perimeter" LRP

4. TELETREATMENT VIA INTERNET USING VTB SOFTWARE

Due to the high prevalence of untreated out-patients in many countries suffering from neurovisual disorders, such as homonymous hemianopia, the inconvenience and costs of treatment sessions on in-patients but increasing world-wide availability of internet access, a pilot study on neurovisual teletreatment has been initiated. To control proper fixation behavior during [2]stimulus presentation (within the treatment area on the screen) **teleproctoring** of the patient's gaze and eye movements via a video monitoring system on top of the PC or TV screen may turn out as an additional low-cost solution to further improve treatment quality.

4.1 System requirements

The predominant issue was creating a simple-to-use teletreatment VTB program version for our out-patients. The actual software (VTB 1.5) is runable on > 386 computers.

Patients have two **options starting the program**

- Program files are installed on the PC the patient is practising with. After a treatment session these files will be send as an Email-attachement (MIME-format) to a computer of the Institute of Medical Psychology in Lubeck. As a file transfer protocol the standard TCP/IP is used. The file transfer is possible with almost every standard Email software
- VTB software is started on our homepage by entering the patient's personal identification number (PIN) und username. The training will be an online-application with or without teleproctoring.

Both modes of treatment initialization require internet acccess.

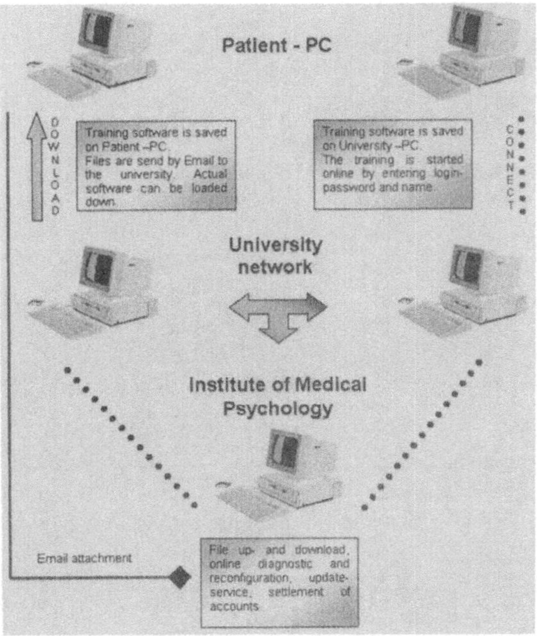

Figure. 2 Neurovisual Teletreatment via the Internet Using VTB Software

4.2 Security

To meet security requirements such an Email has to be crypted by software like "pretty good privacy"[41], a freeware plug in for Email programs like Outlook which ensures a military security level[42]. In both initialization modes, the patient's files will be saved on our PC for diagnostics, interaction and account balance. Those files are protected for unauthorized access by the firewall of the university-communication network[43]. The treatment and program files are very small (< 50 KB program file, < 20 KB training file). The danger of spoofing (manipulation of files) is negligable due to short transfer times, in the order of seconds. Further protection may be realized by separating program and data servers. That way attacks against the program or data server alone wouldn't be dangerous. The "inner security", which means attack-prevention within the local university network, is established by login–passwords and –protocols.

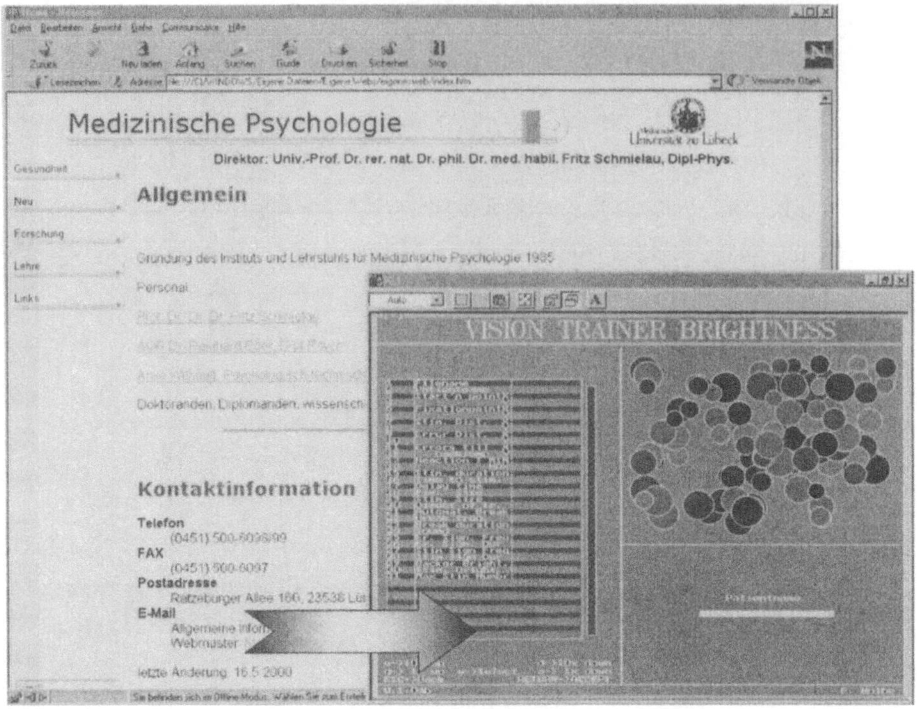

Figure 3. VTB Software start from our homepage

5. CONCLUSION

There are several advantages of VTB neurovisual rehabilitation via the internet:

- There is no mailing delay (**time**)
- All actual treatment files may be evaluated at once, which ensures a much faster response with respect to the patient's rehabilitation progress (**actuality**)
- In case of of treatment software initialization from our homepage, individual updating is easy, since each patient uses his own version uploaded on the homepage.(**individuality**)
- Teletreatment makes in-patient treatment partially obsolete (**comfort** and **low costs**)
- Neurovisual rehabilitation is made available to many out-patients world-wide, who have been left untreated (**rehabilitation potential**)

ACKNOWLEDGEMENTS

Financial support is gratefully acknowledged from Kuratorium ZNS (scientist position), ELWA AG (LRP software development by Thomas Schmielau) and especially from the Medical University Lubeck MUL and the University Clinic Lubeck UKL (extra funding during the last decade)

REFERENCES

1. Vliegen; J., and Koch, H. R., 1974, Die klinische Bedeutung der homonymen Hemianopsie, in : Zihl , J., and von Cramon, D., 1986, Zerebrale Sehstörungen, Kohlhammer, Stuttgart
2. Zihl, J. and von Cramon, D., 1986, Zerebrale Sehstörungen, Kohlhammer, Stuttgart, p 38 Tab .3
3. National Stroke Association, 1998, Acute Ischemic Stroke: Diagnosis, Management and Complications, http://www.geocities.com/HotSprings/4357/facts.htm
4. American Heart Association, 2000, http://www.americanheart.org/statistics/biostats.bioas.htm
5. Heiss, W. D., 1998, Die Vorbeugung, Diagnostik und Therapie des ischämischen Schlaganfalls (Insult), http://www.medicineonline.de/rat/diag/ischschlag.html
6. Medizinfo®Schlaganfall, 1998, Epidemiologie: Rehabilitationsprognose bei Schlaganfall, http://www.medizinfo.de/schlaganfall/epiprog.htm
7. Rossi, P. W., Kheyfets, S., Reding, M.J., 1990, Fresnel prisms improve visual perception in stroke patients with homonymous hemianopia or unilateral visual neglect, Neurology, 40, pp 1597-1599
8. Schmielau, F., 1994, The Lubeck Reaction Time Perimeter-A new device for diagnosis and training of visual defects, in: Mildner (Ed.), Medical Technology Network in the Baltic Region, Mildner, Lubeck, pp 181-189
9. Zihl, J., and von Cramon, D., 1985, Visual field recovery from scotomata in patients with postgeniculate damage. A review of 55 cases. Brain, 108, pp 335-365

10. Balliet, R., Blood, K. M. T., and Bach-y-Rita, P., 1985, Visual field rehabilitation in the cortically blind?, J. Neurol. Neurosurg. Psychiatry, 48, pp 1113-1124

11. Kerkhoff, G,.2000, Neurovisual rehabilitation: Recent developments and future directions, J. Neurol. Neurosurg. Psychiatry, 68, pp 691-706

12. Schmielau, F., 1997, Treatment-induced recovery of visual fields in stroke patients, Society for Neuroscience, 23, 649.11

13. Schmielau, F., Wong, E.K., and Ling, C.A., 1998, Treatment-induced recovery of visual function in hemianopia, Invest. Ophthalmol. Vis. Sciences, 39, 2571-B428

14. Aldrich, M. S., Alessi, A. G., Beck, R. W., and Gilman, S., 1987, Cortical blindness: Etiology, diagnosis and prognosis, Ann. Neurol., 21, pp 149-158

15. Bergman, P.S., 1957, Cerebral blindness, Arch. Neurol. Psychiat., 78, pp 627-658

16. Kommerell, G., 2000, Training of explorative saccadic strategy versus training of visual field, ZNP, 11, pp 86-88

17. Bogousslavsky, J. Regli, F., and van Melle, G., Unilateral occipital infarctation: evaluation of the risks of developing bilateral loss of vision, J. Neurol. Neurosurg. Psychiatry, 46, pp 78-80

18. Poppelreuter, W., 1917, Die Störungen der niederen und höheren Sehleistungen durch Verletzung des Okzipitalhirns, in: Die psychischen Schädigungen durch Kopfschuss im Kriege 1914/16 mit besonderer Berücksichtigung der pathopsychologischen, pädagogischen und sozialen Beziehungen, 1, Voss, Leipzig

19. Poeppel, E., Brinkmann, R., von Cramon, D., and Singer, W., 1978, Association and dissociation of visual functions in a case of bilateral lobe infarctation, Arch. Psychiat. Nervenkr., 225, pp 1-21

20. Zihl, J., von Cramon, D., and Poeppel, E., 1978, Sensorische Rehabilitation bei Patienten mit post-chiasmatischen Sehstörungen, Nervenarzt, 49, pp 101-111

21. Zihl, J. and von Cramon, D., 1979, Restitution of visual function in patients with cerebral blindness, J. Neurol. Neurosurg. Psychiatry, 42, pp 312-322

22. Schmielau, F., and Potthoff, R. D., 1990, Erholung von Sehfunktionen als Folge spezifischer Rehabilitationsverfahren bei hirngeschädigten Patienten, in: Frey, D., Bericht uber den 37. Kongress der Deutschen Gesellschaft fur Psychologie, 1, Hogrefe, Gottingen, pp 123-124

23. Schmielau, F., 1989, Restitution visueller Funktionen bei hirnverletzten Patienten: Effizienz lokalisationsspezifischer sensorischer und sensomotorischer Rehabilitations-massnahmen, in: Jacobi, P. (Ed.) Psychologie in der Neurologie, Springer, Berlin, pp 115-126

24. Wiegmann, U., 1992, Restitutionstherapie von homonymen Gesichtsfelddefekten nach einer Hirnschädigung: 3 Einzelfallstudien, Diploma Thesis, Hamburg,

25. Schmielau, F., 1993, Legal Protection of registered design No. G 93 05 147.6,Germany

26. Schmielau, F., 1997, Patent No. 594 04 213.5-08, Germany

27. Schmielau, F., 1997, Patent No. EP 94 110 2555.0-2305/0 689 822, Europe

28. Schmielau, F., 1996, United States Patent, No. 5,534,953, Date of patent: Jul. 9, 1996, Appl. No. 269,697, filed Jul. 1, 1994

29. Schmielau, F., Training device for the therapy of patients having perception deficits

30. ELWA AG, Gorwihl, Germany

31. Schmielau, T., 1992, LRP Software in: LRP Manual 1998, ELWA AG, Gorwihl, Germany

32. Schmielau, F., 1994, The Lubeck Reaction Time Perimeter-A new device for diagnosis and training of visual defects, in: Mildner (Ed.), Medical Technology Network in the Baltic Region, Mildner, Lubeck

33. Schmielau, F., 1997, Treatment-induced recovery of visual fields in stroke patients, Society for Neuroscience, 23, 649.11

34. Schmielau, F., Wong, E. K., and Ling, C. A., 1998, Treatment-induced recovery of visual function in hemianopia, Invest. Ophthalmol. Vis. Sciences, 39, 2571-B428

35. Kasten, E., 1989, Commodore C 64 Software: SEHTRA, VISURE, FORMTRA, FARBTRA, Lubeck
36. Kasten, E., 1993, Computergestutzte Rehabilitation bei cerebral verursachten Sehstorungen, Ph D Thesis, Halle
37. Kasten, E., Behrens-Baumann, W., and Sabel, B. S., 1998, Computer-based training for the treatment of partial blindness, Nature Medicine, 4, pp 1083-1087
38. NovaVision VRT, Medinova Medical Cons GmbH, Magdeburg, Germany
39. Schmielau, F., 2000, The Efficiency of Visual Rehabilitation, ZNP, 11, pp 89-90
40. Schmielau, F. (unpublished results)
41. http://www.pgpi.com
42. Schneider, B., 1994, Applied Cryptography, Wiley & Sons, New York
43. Cheswick, W. R, and Bellovin, S. N., 1996, Firewalls und Sicherheit im Internet", German of Maus, Addison-Wesley, Bonn

Knowledge Management in the Field of Radiology

DIRK KRECHEL

Cyclops Project, Knowledge-Based Systems Group, University of Kaiserslautern, Erwin Schrödinger Straße, 67663 Kaiserslautern, krechel@informatik.uni-kl.de

Key words: knowledge management, workflow management, DICOM, image processing, CBR

Abstract: The medical knowledge for the diagnosis of radiological images is growing very fast. For the medical practitioner in private radiological hospitals this is a problem, because of the few number of specialists and the high fluctuation of the medical staff. Knowledge management techniques used in business applications may help to save this knowledge and share it between all radiologists in the organisation. In this paper a concept of a small OMIS (Organisational Memory Information System) for private radiological practices is presented. The system is based on a workflow engine and a knowledge based image processing system.

1. INTRODUCTION

Medical knowledge grows exponentially. One future goal for medical informatics should be to improve patient care by utilising this knowledge. Commercial applications of knowledge management methods show high benefits for companies that invest in this field. These methods should be applied to the medical domain.

Actual developments like the IHE initiative of the HIMSS & RSNA try to integrate all kind of information systems in health care organisations. But this initiative is focused on one kind of knowledge, the organisational knowledge of the health care enterprises. It's focused on the process improvement of the daily administrative workflow. But a second kind of knowledge should also be managed, the medical knowledge, that consists of the general textbook knowledge and the personal experiences of the medical staff. To share this knowledge between all medical practitioners is a second important field of research. There are some efforts in creating medical guidelines, protocols and to build some basic medical ontology's, like UMLS or the SNOMED thesaurus.

In the German-Brazilian cooperation project Cyclops we try to build an integrated solution for both knowledge types. A process-oriented view is used for the knowledge representation. The main component is a workflow system based on the standard model of the workflow coalition and the process definition in the IHE Technical Framework. The key technologies used for knowledge management are Case Based Reasoning, knowledge based image processing and business process models. The system is developed in close cooperation with private radiological hospitals in Germany and Brazil. One aim is a easy knowledge exchange between the medical partners in both countries.

2. KNOWLEDGE MANAGEMENT

The aim of knowledge management is the optimal use of the resource knowledge in an organisation. The organisational knowledge is stored in documents, procedures, products and in the 'heads' of the staff.

Research projects in computer science related to knowledge management can be classified in two general types:

- Process oriented
- Product oriented

Process oriented works mainly use basic techniques like CSCW (Computer Supported Cooperative Work) and Workflow Management. The product oriented systems use methods from the fields document management and knowledge based information systems.

In this section we will explain some basic concepts of knowledge management systems

2.1 Organisational Memory Information System (OMIS)

A OMIS[1] system is the integration of basic techniques into a computer system, that is collecting, structuring and updating all kinds of organisational knowledge and that, active presents all relevant information that is needed to improve specific work tasks.

An OMIS has three main components for:

- Knowledge acquisition and maintenance: Knowledge Acquisition and maintenance are the most critical points for the success of an OMIS. The effort can only be decreased by good acquisition tools and semiautomatic knowledge discovering methods (e.g. text mining).
- Integrated knowledge access and presentation: One goal of an OMIS is an effective and standardised access to the knowledge memory.
- Fast and easy knowledge search and reuse: According to the degree of formalisation in OMIS, different processing techniques for knowledge and meta-data are needed. Expert systems are suitable very formalised knowledge and Cased Based Reasoning techniques can handle the less structured types

2.2 Knowledge Representation

What is knowledge and how can it be represented in an information system? If we talk about knowledge we must distinguish between explicit and implicit knowledge. Explicit knowledge is present in things like rules, norms, laws, procedures, recommendations, guidelines, hierarchical relationships, properties or conditions. These knowledge sources can easily be transformed in a computer representation. To capture the implicit knowledge that is hidden in data or in the heads of domain experts is very hard. Humans acquire knowledge by actively creating and organising their own experiences.

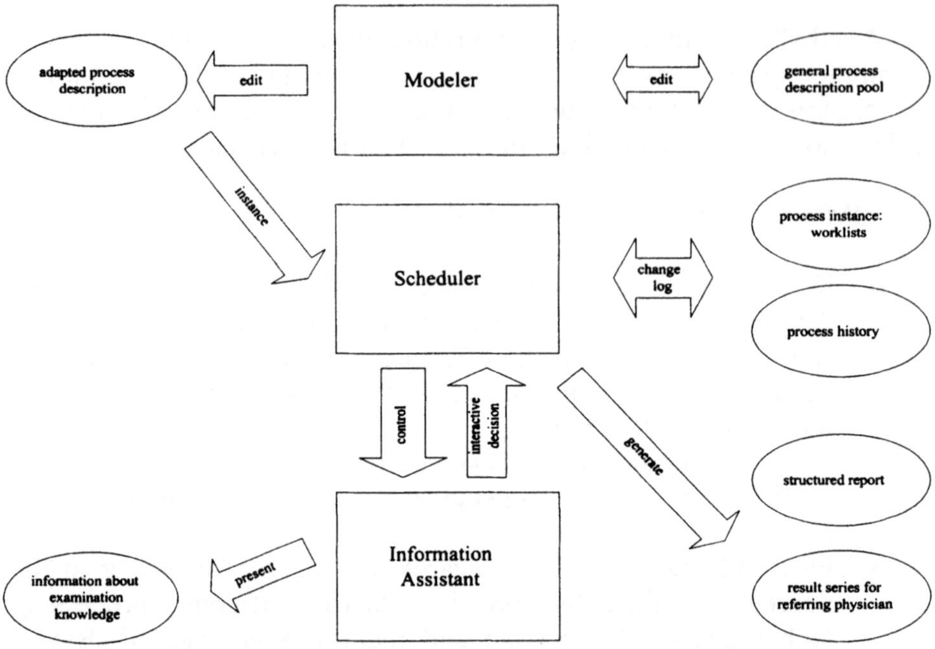

Figure 1: Architecture of the Radiology Assistant knowledge management system

The second problem is the knowledge representation language. Often there is a conceptual and terminological confusion if experts talk about domain knowledge. A shared understanding of terms and concepts is needed to build a knowledge management system that is used in an organisation. Ontology[2] is the term used in computer science for a shared conceptualisation. A well defined ontology is the semantic base of a knowledge management system. It consists of concepts, relations and rules that build the consensus of the group of people in a specific working area.

2.3 Current developments in the medical domain

In the medical domain there are several developments that can be very useful for building knowledge management systems.

One, that is based on the idea of an integrated solution for the management of the administrative workflow of a radiological department, is the IHE, a joint initiative of HIMSS and RSNA, that was founded to stimulate integration of information and imaging systems and other software components and resources for healthcare by encouraging existing standards like DICOM and HL7. IHE shows the architecture of an integrated system, which connects several components of a hospital like RIS, modalities and PACS, so that e.g. redundancy of data and function can be reduced and existing processes can be improved. As a result, examination times shall be accelerated and the cost of an examination shall be reduced. IHE defines a Basic Process Flow Chart consisting of 18 steps, starting with the registration of a patient, and including steps like the arrival of the patient in the hospital, the taking of the images and the reporting of the results.

The advantages of the process oriented approach are described by Th. Wendler[3] in his paper in "Applied Radiology". One of the key sentences in this paper is: "*Introducing information technology as such, will not automatically improve processes. Current IT solutions in radiology are predominantly data-focused. For the challenges of the future we need process integration in addition to data integration.*"

A lot of interesting work is done in the field of medical ontologies. Some examples are the Unified Medical Language System UMLS[4] and the Systematised Nomenclature of Medicine (SNOMED) .

Research on process oriented guidelines for medical examinations was made by a group in italy[5]. They used the commercial workflow features of Oracle.

3. RADIOLOGY ASSISTANT

The Radiology Assistant[6] is a knowledge management tool for private radiological practices. We try to improve the daily work of our medical partners by the integration of our tool that should act as the Organisational Memory of the practice. We have chosen a process oriented approach[6].

3.1 System Architecture

Figure 1 is showing the system architecture of the Radiology Assistant. It is very similar to the reference model of the workflow coalition. The system has three main components that we will discuss in the next sections.

3.1.1 The Modeller

The modeller is our knowledge acquisition tool. This tool has three components. The first component is for the acquisition of the process models and the integration of extern tools in the process enactment. The second component is for storing and retrieving example cases for a specific examination. The third component is an interactive tool to create and test image-processing sequences that help the physician to perform the diagnosis steps.

The modeller is used to define the process descriptions. We build different kinds of process descriptions. First processes that presents the knowledge about specific examinations. This medical knowledge tasks, consists of worklists with the standard examination steps, textual descriptions, example cases and specific image processing routines that can help to find suspicious regions in big data-sets.

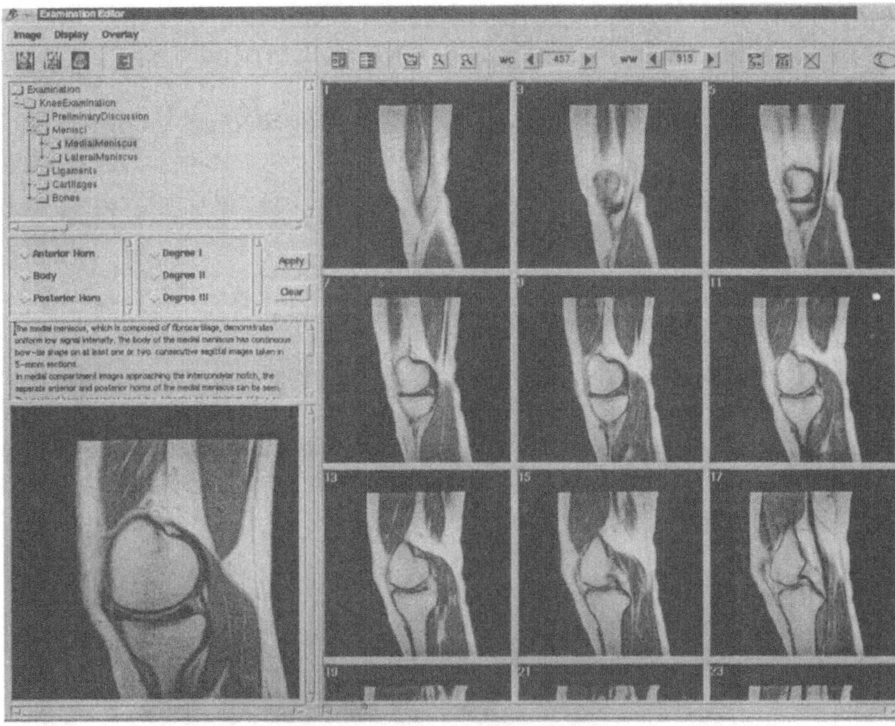

Figure 2: Graphical user interface of the information assistant

Second we are building a process model of the organisational workflow in our partner hospitals and plan to integrate the existing information systems. There must be some benefits for the users in the routine work. This is essential for the acceptance of the system.. The DICOM worklist-management helps to integrate the system in the existing RIS.

3.1.2 The Scheduler

The Scheduler is our workflow engine. It consists of two components. The first is a simple dataflow machine similar to the petri nets. And the second component is a knowledge based image processing system[7] that is based on goal-task driven configuration system for automatic composition and execution of image processing pipelines. The scheduler generates worklists for the information assistant and controls the process execution. If decisions are made interactive the scheduler waits for specific inputs from the information assistant. The process execution is logged in a Truth Maintenance System.

3.1.3 The Information Assistant

To support the physician during his daily work we developed an information assistant The information assistant is integrated in our DICOM viewer and the new standard parts presentation state and structured reporting help us to exchange and present the knowledge.

Figure 2. is showing a screen shot for a typical examination in our partner practises. On the left side of the window typical examination steps for a knee are displayed. The radiologists can expand the single steps and mark findings on the overlays of the slides. For all anatomical parts that could be damaged typical pathologies are displayed and can be selected for the automatic report generation. For all possible pathology a textual description and example pictures are displayed. The system switches to the best series for finding the searched pathology.

4. CONCLUSION

We are currently testing our system in one of our partner practises in Mainz. The first modelled examination is a typical knee MR protocol.

The acquisition tool will help our partners to build and maintain a hospital experience base, that represents the best practice of their medical staff. A medical student together with the experienced radiologists is currently using the modeller to generate several examinations. Interesting example cases can be stored directly during the work. We hope that the system will be accepted in the routine work and improve the quality of the practise.

Our simple system shows the advantages of knowledge management methods in medicine. It is a small example for a OMIS in radiology, but it is only a small step and a lot of work has to be done in this field.

REFERENCES

1. Abeker A. and Decker S., 1999, Organisational Memory: Knowledge Acquisition, Integration and Retrieval Issues. In *Proceedings of the 5th German Conf. On Knowledge based Systems* (Puppe ed.), LNAI vol. 1570 Springer Press Heidelberg and Berlin.
2. Gruber T.R., 1993, A Translation Approach to Portable Ontology Specifications. *Knowledge Acquisition* 5 (2)
3. Wendler Th.,1999, Workflow management – A perspective for more efficient radiology, *Applied Radiology December99* .
4. U.S. Department of Health and Human Services and National Library of Medicine, 2000, *UnifiedMedical Language System – UMLS Knowledge Sources*. 11th edition.
5. Quaglini S., Mossa C., Fassino C., Stefanelli M., Cavalinni A., and Micieli G., 1999, Guidlines-Based Workflow Systems *AIMDM 99* (W. Horn at al., eds.) Springer Press Heidelberg and Berlin , pp. 65-75.

6. Dirk Krechel, Faber Kerstin, Reidenbach Daniel, Blasinger Klaus, von Wangenheim Aldo, Comunello Eros: A Workflow Component for Knowledge Managment in the Field of Radiological Examinations, *Bildverarbeitung fuer die Medizin 2000*, München 2000, Springer Informatik aktuell ISBN 3-540-67123-4

7. Dirk Krechel, Christian Pape, Michael M. Richter, Wissensbasierte Bilddeutung in der Medizin am Beispiel des CYLCOPS-Systems, *Angewandte Mathematik, insbesondere Informatik (Hrsg. Patrick Horster,* Vieweg-Verlag, ISBN 3-528-05720-3

PLATFORMS AND MEDICAL NETWORKS

The Telemedicine Project of the Medical Service of the Bundeswehr

C. OTTO
Department of Medical Informatics, Medical Office of the Bundeswehr, Platanenweg 29, 53225 Bonn

Key words: Quality improvement of medical treatment, commercial-off-the-shelf-products, telematic procedures, communication standards, compatibility, interoperability.

Abstract: The maxim governing the accomplishment of medical tasks by the Medical Service of the Bundeswehr stipulates that the soldiers will, in case of illness, accident or injury, receive medical treatment to the standard they can expect in the Federal Republic of Germany. In this context the introduction of telemedical applications into the Medical Service will improve the quality of the medical care through better and faster access for the medical officer assigned to a unit, a ship or within theatre medicine on missions out-of-area for requesting support to expert knowledge from military facilities or the civilian sector. The telemedical system is based on a high-class commercial PC which can be extended as needed with medical devices and additionally provides access to medical data bases through web-technology. The 54 participants of this telemedical network communicate via EURO-ISDN lines with sufficient bandwidth. If terrestrial lines are not available, e.g. from vessels of the Navy and from the field hospitals out-of-area, telematic procedures will be performed on satellite capacity. This future-orientated telematic approach with its modular upwards-compatibility will provide - by using communication standards - an improvement of medical quality through additional expert-advice, civil-military co-operation and the interoperability between medical services of allied armed forces on common missions outside of Germany.

1. INTRODUCTION

Telemedicine is the investigation, monitoring and management of patients and the education of patients and staff using sytems which allow ready access to expert advice and patient information no matter where the patient or the relevant information is located.

Health telematics may be defined as " the applying of modern **telecommunications and information technology** within the **health service ".** Hence, telemedicine is the using of health telematics for bridging the distance between the patient and the doctor or between several attending doctors. A prerequisite is the existence of telecommunications technology/IT infrastructure and digital medical data and information.

Possible telemedical applications include:

- **Teleconsultation** between the requester and the expert on a first-opinion basis and on a second-opinion basis
- **Cooperative care** (between ambulant care and hospital)
- **Clinical teleconference** (e.g. X-ray picture conference)
- **Teleteaching** (needs to be organized !)
- **Telecare** (telemonitoring of patients with chronic diseases, e.g. cardiac insufficiency, SIDS)
- **Remote archiving** (theater of operations - *Bundeswehr* station hospital - PACS)

2. PAST ACTIVITIES

Experience of using telemedical systems in the *Bundeswehr* medical service stretches back to the year 1993. The neurosurgery ward of the *Bundeswehr* station hospital in Ulm has since received from surrounding hospitals computer tomograms and X-ray images of patients with neurotraumas that are transmitted using a **photophone system**. The decision to perform a neurosurgical operation at the *Bundeswehr* station hospital in Ulm after the patient has been transported there by air, or to provide conservative therapy at the hospital in question, can thus be taken on the telephone, or via the Euro-ISDN link, which has been in use for three years now.

In late 1996, a **high-end teleradiology/videoconferencing system** was implemented between the Medical Center in Bonn and the radiology ward of the Koblenz-based *Bundeswehr* Central Hospital. The system is praised for permitting loss-free transmission from Bonn of digitized X-ray pictures. Following transmission, all X-ray pictures produced in Bonn by unit surgeons or specialists can be displayed and discussed, as required, in on-line teleconferences, using interactive graphical elements, before they are archived in the Koblenz-based PACS-RIS system, which occurs after a report that has been sent to the requesting surgeon/General Practitioner.

The two just-mentioned projects are specialized ones which, owing to the **high costs involved,** their **restricted applicability** and **expensive operation,** cannot generally be used in everyday unit medical care, that is in general medical care. Moreover, the system in Bonn will be cannibalized this summer, which means that it will be reinstalled at the *Bundeswehr* station hospital in Ulm, featuring as functionalities picture evaluation and the telemedicine PACS of the MAGIC STORE system.

3. IMPLEMENTATION OF TELEMEDICINE

In April 1996, the Department of Medical Informatics of the *Bundeswehr* Medical Office was tasked to develop - in a mixed working group led by the Medical Office and assisted by representatives from the medical services of the Army, Air Force and Navy and from other *Bundeswehr* agencies - a concept for **extensively operating low-end telemedicine workstations** at home in Germany.

Because the ***Bundeswehr*** **lacked expert knowledge in the novel field of telematics,** the Cologne-based institute of aerospace medicine of the German Aerospace Center (DLR) was selected as a competent partner in the study. In cooperation with a mixed study group of the medical service, the organizational, legal, technical and conceptual parameters for extensively introducing telemedicine have been defined and implemented in keeping touch with the progress of the project since November 1997.

3.1 Telemedical Workstation Requirements

Initial considerations of using telemedicine on an extensive basis have resulted in the following requirements:

- Peacetime coverage of the **requests of both unit surgeons and specialists** with a wide range of technical applicability
- Use of " commercial-off-the-shelf-products "
- Consideration of IT and communication **standards**
- User-friendly operation and technically reliable systems
- Main efforts of support: **teleconsultation in the form of second opinion**
- Teleradiology support provided only where required medically and acceptable economically
- **Internet access** for e-mail communication and use of technical data bases of the World Wide Web
- **Office communication** support

3.2 Telemedical Workstation

The schematic telemedical workstation shown in the graph consists of commercial IT and medicotechnical components. It can be expanded on a modular basis so as to include additional functionalities, such as image pickup, processing, digitizing and communication.

The basic module is an IT-standard workstation complete with printer, incident-light scanner and multimedia equipment. Via a **video changeover switch** for S-VHS and FBAS signals it can be linked to medicotechnical modalities, such as a **plain-survey camera with a cold-light lamp mount**, which is used for dermatoscopy purposes. It can be linked to existing **peripheral equipment, such as sonographic equipment plus video recorder** (endoscopy) and special camera systems, such as otoscopy/laryngoscopy systems, which currently are being tested by otolaryngologists in Ulm and Kempten. Other modular links are being examined within the scope of clinical validation. Beginning in the autumn of 2000, a certified crypto-box will encrypt not only DICOM-3 communication but also videoconferencing with its file transfer and Whiteboard functionalities.

Where it is possible to fall back on terrestrial links, **communication** is ensured by using two linked B channels of a EURO-ISDN link. The resulting transmission capacity of 128 Kbit/s permits H. 320-standard videoconferencing, with the quality of video telephony being limited yet acceptable.

Communication software permits transmission of high-resolution still images for diagnosis purposes. Where there are no terrestrial links available, the means to use is satellite communication. During the Kosovo mission, for instance, use is being made of *Bundeswehr*-owned satellite capacities (C band ground stations, n x 64 Kbit/s) of adequate bandwidth, or, on board Navy ships, of civil providers, such as InMarSat B systems (64 Kbit/s).

3.3 Functionalities

Supported functionalities include videoconferencing, file transfer, Whiteboard and image processing. The Rajlovac-based field hospital in Bosnia and the Prizren-based field hospital in Kosovo, which have been integrated within the telemedicine network since November 1998 and February 2000, respectively, provide data transfer via the DICOM-3 interface. A fluoroscopic-image scanner permits digitizing conventional X-ray images. These images can be displayed on the control PC before they are transmitted as compressed JPEG images to the experts in Koblenz and Ulm lossy or lossless and encrypted, using the DICOM-3 format. In Koblenz and Ulm the X-ray-images can be displayed, evaluated, and archived in the PACS-RIS system. The agencies there require supplementary hardware and software configurations with routers and hubs when it comes to using TCP/IP standard.

4. INTEROPERABILITY AND STANDARDISATION

Combined missions with alliied forces and the need to co-operate with civil hospitals like in a host-nation-support underline the demand to use common technical standards in telecommunication in order to enable interoperability und thus maximum synergy of services. Examples for medical data exchange formats are the station hospital based HL-7 standard and the DICOM-3 standard for the transmission and storage of digitalized X-ray images, CT and MRT scans. Communication standards are the ISDN-based H.320 or TCPIP-based H.323 for a larger bandwidth. Common technical plattforms make use of high-class commercial PCs and ISDN-links with sufficient capacity and the EURO-transmission protocol.

5. TELEMEDICAL GUIDELINE

The medical support that military personnel are to receive on operations, and whose standard must correspond to specialist care, is governed by the so called **Einbeck Recommendations of the German Society for Medical Law**. For instance, when it comes to telesurgical procedures, the assumption is that a competent surgeon is available within 10 minutes. In this context, telemedicine cannot be substituted for the specialist, but in the form of telepresence in the operating room can **effectively support the treatment in question by providing a firm second opinion**. In the future, this will be a requirement at the Naval Rescue Coordination Center on board the Combat Support Ship. In any case responsibility will rest with the locally attending physician. He must thoroughly inform his patient about the application of telemedicine and will have the patient acknowledge in writing receipt of this instruction. Telemedicine is not able to substitute operational personnel or to cut responsibility for the treated patients. The employment of telemedicine improves the quality of medical treatment especially in cases of theater medicine and on board of naval vessels. Prerequisite for the responsable use of telemedical systems are clinical evaluations and guidelines.

6. DEVELOPMENT OF OBJECTIVES

Initial conceptual considerations focus on **supporting the unit surgeon** through requester-expert teleconsultation and using the Internet. The **objective** is to improve the quality of medical treatment and to cut costs by limiting referrals to the civil sector and transportation to *Bundeswehr* station hospitals/specialist medical centers. When the Rajlovac field hospital was equipped in November 1998, the **increasing importance of medical support on operations** was reflected in an **improved information transfer** between the theater of operations and the home country and in providing support where **medical evacuation** is indicated. In the face of increasing personnel cuts in minor clinical specialties, the **overall objective of telemedicine is to improve the quality of results of medical support and treatment on operations**. The telemedical workstation, which must be operated by skilled personnel to transmit relevant findings to the nearest expert station (field hospital level) and to discuss such findings, is particularly suitable in situations where medical support on operations is limited. The idea is to **provide second-opinion**

consultation and/or to recommend the referral of the servicemember in question, as required.

In mid-1998 the first eight telemedical workstations were successfully implemented at the medical agencies of the various services and within the organizational component known as the Central Medical Agencies of the *Bundeswehr*.

In terms of frequent teleconsultation, teleconferences and arranged teleteaching, the pilot test progresses successfully where people know each other (human factor!). We will take account of this fact - also by holding regular user meetings - when it comes to providing a vertical link between the organizational medical service and the expert stations at the Central Medical Agencies of the *Bundeswehr*, so as to regionally assign the medical agencies to "their *Bundeswehr* station hospitals and specialist medical centers."

Beginning in the early summer of 2000, four *Bundeswehr* station hospitals and four specialist medical centers of the Central Medical Agencies of the *Bundeswehr* will act as expert stations, and two field hospitals, one rescue center for reaction forces, and 33 medical centers of the Army, Air Force and Navy will act as requesting medical agencies.

7. RESULTS OF TELEMEDICAL SUPPORT ON OPERATIONS:

Telemedicine within the SFOR and KFOR contingents is aimed at providing **optimum medical support** for our servicemembers in an **adverse operational environment**.

Telemedical applications meanwhile established include teleconsultation, X-ray image conferences, remote archiving, and the initial stages of teleteaching. Internet access via single-user computers has been implemented as well by now.

The following results can be stated:

- Using a **uniform telemedical standard workstation** that the telemedicine user and the telemedical workstation operations manager know from their home station has proven worthwhile and is increasingly gaining acceptance.

- **Teleconsultation calls for the availability of experts,** in particular when it comes to videoconferencing, but temporary duty has the number of experts decreasing.

- **Operating concepts and adjusted IT security concepts** facilitate and ensure the use of the telemedical workstation. This includes supervision by higher agencies. Identifying telemedicine tasks at least three months prior to temporary duty and providing pertinent training have proven worthwhile.

- **Remote archiving of digitized X-ray images** following clinical radiology validation requires the procuring of X-ray image evaluation stations and modern PACS systems for storing and archiving, and the horizontal networking of the four major *Bundeswehr* station hospitals at home, which needs to occur on both a cryptographically secure and successive basis. Beginning in 2001, the procurement of one HOWTEC X-ray scanner each for the Berlin and Hamburg *Bundeswehr* station hospitals will facilitate teleradiological expert communication.

- **Clinical validation**, the procedures and methodology of which were put into force by the Commander of the *Bundeswehr* Medical Office in his 31 March 2000 Directive, can and must **improve and expand the range of applicability of telemedicine also on operations.**

8. CLINICAL VALIDATION OF TELEMEDICINE

Clinical validation of telemedical applications is **a form of quality management** regarding methodical application of telemedical procedures in a scientific comparison with conventional medical methods, the gold standard.

The clinival validation will be enforced in the Central Medical Services of the Bundeswehr under responsibility of the chairmen of the clinical consulting groups.

In terms of application of telemedical procedures, it has to be determined on a binding basis what is

- **functionally desirable,**
- **technically feasible,**
- and **legally permitted.**

It is in the nature of things that the most significant progress in terms of advancement of validation is seen in those functional areas which are based on imagery information and pictorial documentation, namely radiology, (neuro)surgery, dermatology, otolaryngology, ophthalmology, laboratory and transfusion medicine, internal medicine and dentistry.

9. POTENTIAL FOR IMPROVEMENT

- **With the planned extensive introduction of IT equipment** into the KIS, and above all into the Information system for our GP's, and with the planned integration of telemedicine (secured by firewall systems and including network-based solutions on the basis of existing IT infrastructure/LAN links of the agency in question) the applicability of the **telematic methods,** which have already been defined by our "telemedicine" project and which have already **proven their worth in clinical operations,** can be enhanced by **integrating providers on a broader basis** (telematic data are submitted to the expert and can be properly displayed; the expert does not need to go through the building for hundreds of meters to pick them up).

- There is a need for **more substantial ADP source documents,** because the current training and teaching capacities of the German Aerospace Center and of the *Bundeswehr* Medical Academy are nearing their limits (alternatives: civil on-site courses, multipliers).

- The **planned extensive implementation of a *Bundeswehr* Intranet** will support off-line/ asynchronous teleconsultation and telearchiving of digitized X-ray images.

- **Synergistic effects** may be enhanced **by increased cooperation among all those involved.**

- **Telemedicine takes time and requires commitment**; to the users the decisive source of motivation is acceptance, which may be increased if attention is paid to user problems and to solving them using telematics methods - in other words, what matters is implementing solutions relevant to practice.

- **Operating concepts and documentation concepts** must permit telemedicine to develop in the peripheral sphere on an orderly basis.

10. CURRENT AND FUTURE DEVELOPMENTS:

- The **roll-out** and the **integration of the 38 new telemedical workstations** in the first quarter of 2000 can now be considered successfully completed, except for a few lingering problems in terms of communication technology.

- Another subproject is the **implementation of telemedical systems on board of Navy vessels.** Beginning in September 2000, there will first be a communication test and a pilot installation on board the Frigate 123, **Mecklenburg-Vorpommern.** This installation must be geared toward the existing on-board infrastructure in the unspacious dispensary and toward the telemedical workstation's internal (inside the ship) and external communication facilities. This also requires the operation of mobile systems at the **Naval Rescue Cooordination Center.** Beginning in the autumn of 2000, implementation will occur on board the **Operation Support Ship "Berlin"** at the **container-based modular medical facilities.** Transmission of high-resolution still images and video sequences from the operating room, transmission of digitized X-ray images, microscopy images and ultrasonic images, and transmission of biometric data from the ECG room and from the clinical laboratory to a central Navy expert station at the Naval Medical Institute at Kiel, to include forwarding of such data, as required, are at the center of future planning, as is the integration of a basic module of the hospital information system.

- In autumn, during PfP exercise **CoLa** 2000, tests will be conducted using Internet Protocol Version 6 (ipv 6) and Standard H. 323 VK and a satellite link of more than 2 Mbits bandwidth between a telemedical workstation in southern France and the expert station at the *Bundeswehr* Central Hospital in Koblenz.

- Clinical validation has already been addressed.

- An **optional master contract** concluded with an IT company established on the market, along with support provided by the German Aerospace Center as a consulting partner until 2003, will ensure the flexible procurement of the future telemedical equipment for the medical service. For instance, two special workstations for the Navy and another 35 land-based telemedical workstations will be fielded in the autumn of 2000.

Bottom Line: (Retired) Brigadier General Russ Zajtchuk from the DoD Telemedicine Testbed at Fort Lee, VA, puts it as follows: "In the long run, progress in telemedicine will be most rewarding if it leads to an improvement of costs, quality and access to treatment..."

Integrated Health Care Within the Compound Network Bottrop

[1]R. FEINEN , [1]H. SCHMITZ, [2]B. REHBERG AND [2]D. KEITE
1) Laufenberg GmbH & Co KG Bochum;
2) SerCon GmbH Mainz

Abstract: Networked based acting and thinking requires a new intelligent IT quality standard. Based on a common information-platform, the planning, documentation and quality of patient care management ought to be co-ordinated. The common information-platform should co-ordinate the treatment of in- and outpatients and must ensure the communication transaction. So far in theory: In practice a lot of questions regarding data security in complex network solutions has to be answered. In addition the demand of the bill GKV-reform 2000, which regulates integrated health services in its § 140, must be considered.
How can such an IT-concept be realised? Laufenberg GmbH & CoKG together with SerCon GmbH developed a software solution called NETSIGHT. This solution is based on the electronic patient record of the Hospital Information System PROSIGHT® and LOTUS NOTES®. These systems are widespread and approved and due to that aspect the investment is ensured and planning reliability is given for the future. Due to the top priority of data security the Groupware System LOTUS NOTES R5 was selected. NETSIGHT is a comprehensive communication-platform, with an integrated Electronic Patient Record. Yet this product connects hospitals and physicians, within the scope of a pilot project of the Bundesknappschaft. The result is an integrated healthcare-platform.abstract.

1. INTRODUCTION

"The german health system impends an infarct", said DER SPIEGEL (44/99) a few months ago. And in fact the situation demonstrates dramatical. The demographic development, increasing mean age and as a consequence thereof an raise of confirmed diseases

Telemedicine: Medicine and Communication
Edited by T. M. Buzug *et al.*, Kluwer Academic/Plenum Publishers, 2001

lead in conjunction with the demand on high class, state of the art medical treatment to a raise of expenses for healthcare, labor market related loss in premium impair the income situation of the health insurances. Thereby previously unknown problems are created in health care. Germans get x-rayed twice as much as a Dutchman ,consult a physician three times more as a Swede and take twice as much medication as Norwegians.

The matter therefore is less that the Germans are sick more often then others. Rather the missing of a healthy co-operation and communication between the different medical attendances. Hospitals, settled physicians and rehabilitation work separately, a result of medical division of work which has grown in decades. There are hardly any control mechanisms or transparency of costs, but inefficiency and inadequate resource exhaustion. The consequence: increasing pressure of cost.

2. THE TREND

Approaches for a solution are for example Managed Care, Case-Management and Evidence Based Medicine underlying an improved interlocking of the medical sector.

Politics, self-governing bodies, expert groups as well as citizens more and more realise that the challenge can only succeed by (means of) interlocking all sectors of the heath care system. Solely thereby a patient can be treated efficiently and an explosion of costs in health care can be embanked. The legal regulations are established currently. Nevertheless, there is a need to change mind. Which has been already realised by the first. More and more practice network coming to existence aspire an information exchange. The aim is next to the quality of processes and of treatment to optimise costs as well.

3. THE CONSEQUENCE

Using a network and consideration of health care require a new quality of IT- Technology. About an information platform as a common basis treatment planning, -documentation and -quality has to be co-ordinated. This information platform needs to be integrated into an EDV technical total concept, interlock ambulatory and stationary treatment and guaranty save communication ways. A functioning and efficient network does only make sense if the junctions exist and are applicative to this web and than actually to spin the web and vitalise it.

Thus clarifies that providing health care institutions with an suitable institution related and over institutional information system which is functioning as a basis for the health care net is a critical factor of success for the further development of the German health care system.

Factor of success for this information system is in turn the possibility of a patient-related data exchange regarding to the requirements of data protection and the accompanied building of mutual trust for all net participants.

Theory so far. The questioning in practice concerning data protection for an over institutional network represents extreme complex. Requirements of § 140 which rules the integrated provision have to be complied. This topic has been discussed much, but how can this IT concept be realised?

4. PROJECT MANAGEMENT

At first the organisational prerequisites have to be accomplished for the establishment of a net. These have already been discussed in this and other meetings. Special attention is to be directed on the strategic orientation of the net (required contracts, inclusion of hospitals, size of the network, network supply), on structure- and procedure organisation (network office, network co-ordination) and on an result-oriented analysis concept (medical and health-economically relevant data management).

Net participants within the integrated provision have at first to be conscious about the personal need for information and the available information.

First of all the considering preparatory concept work needs to be accomplished. What kind of information connected to which patient is to be listed in a network document. And in which organisational connection, in which patient context and which patient-specific information of the electronic network document ought to be accused?

On the basis of need for information and availability of information a soluting concept can be created which has to consider data security and aspects of data security. This information underlying an appropriate communication concept can be implemented. At this point storage of data (central Hosting, ASP (Application service Providing) peripheral Hosting) as well as server- and network administration has to be reflected. Subsequently, implementation and parameterisation of the system are executed.

5. COMMUNICATION PLATFORM

A health care communication platform with an integrated central electronic patient document is necessary. NETSIGHT connects already today hospitals and established physicians within a pilot project of the Bundesknappschaft. The result: an integrated health platform.

NETSIGHT adopts the function of a network server - with the electronic patient document as trick and pivot. Further elements are network-internal Intranet functions such as a forum for discussion, black boards, positive lists of medicines, supply of network-internal treatment guidelines, integration of treatment-specific Internet links as well as patient accounts for the cost-computational analysis.

6. THE NETWORK ORGANISATION

In order to build a ambulatory- and stationary sector wide network, the organisational and legal data security conditions have to meet the requirements.

This must be ensured on the one hand by a central patient data bank and administration of the participating physicians and hospitals, on the other hand by asymmetrical encoding procedures for the data communication and differentiated access authorisations for all defined network participants. Among other functions the explicit patient participation with unlocking the entries belong into the electronic patient document. This means a clearly structured access right administration providing the highest possible transparency of the patient treatment.

7. THE DOCUMENT ORGANISATION

The largest problem with networking of established physicians in-between and with hospitals are the interfaces. The single systems in the practices are rarely connected to each other or connected to a hospital for exchanging Treatment data, since no common communication standard exists. The only interface which at least equals in principle within all practice systems is the BDT (Behandlungsdatenträger-) standard. These usually enable the data transfer between practice systems while changing a system. As for each standard the individual software producer maintain their programming differently near to the standard, thus different qualities of data sets from different systems result.

NETSIGHT enables now to adjust documents of a patient as BDT file into the central electronic patient document and to provide the according information. Special interest was investigated to a complete independence from the respective practice system of the established physicians and from the hospital information system of the hospitals involved. A single prerequisite is the ability, to import or export documents in the so-called BDT format. In the context of this communication a BDT filter process already verifies the data before the dispatch, verifies whether the physician is entitled to transfer these data of this patient into the network document and for whom the document is visible and accessible. Thus an unauthorised data transfer is already prevented before the dispatch. Integrity and authenticity and thus the protection from unauthorised modifications of documents are ensured by appropriate access protection and data security mechanisms.

The query can take place according to the authorisation of the involved directly on a NETSIGHT Client on practice- or hospital computer. Beside text documents, picture documents (e.g. radiographs in the DICOM format), charts (e.g. EKG curves), sound documents (e.g. digital dictation) or scanned documents (e.g. older documents) can be represented.

Additionally it is possible to execute patient-referred documentations within the electronic network document. For the structured, dynamic documentation within the electronic network document a tool for designing non-standard documentation surfaces is supplied. The tool assists within the network to meet the different medical documentation needs. An example of this procedure is the Diabcare form, we are able to provide within the network. On the basis of these dynamic data base entries differentiated analyses naturally possible in the sense of a medical evaluation by standard tools.

8. THE NETWORK COMMUNICATION

The patient-unspecific communication of the network participants is supported by NETSIGHT as well. Scheduling for the private and network-internal date administration, e-mail functions using mailing lists, information forums for network-internal exchange of views, discussion forums with use of Internet links of e.g. on interesting publications, network-internal listings for medical treatment, black boards for messages to all network participants, discussion- and expert forums are available in the Intranet for a permanent information exchange over all sectors of healthcare and regardless any boundaries.

9. WORK-SHARING TREATMENT

A health net develops its largest use for all network participants, if all performances for a patient can be settled within this network on the basis of a co-ordinated procedure of the network participants. Therefore NETSIGHT will support work sharing treatment organisation on the basis of network-internal or external guidelines and thus make in the sense of a Case management Tool for the providing institution wide organisation of the patients treatment available. Individual functionalities for system internal application are:

- Creation of treatment standards.
- Creation of individual treatment plans on the basis of treatment standards.
- Timing and planning of practice -, outpatient clinic -, hospital -, and rehabilitation contacts and procedures on the basis of the individual treatment plan.
- Summing and cumulating from partial plans to individual treatment plan.
- Adjustment of the treatment plan.

CONCLUSION

The common electronic patient document ensures a current transparency at any time of the patients treatment and offers thereby to the physicians, patients and hospitals involved a clear plus at security of treatment.

The co-ordination of the treatment procedure becomes more efficient, repeated examinations can be avoided, medication, primary examinations and anamnesis of the patient presents transparent, the quality of treatment increases. At the same time by proving Intranet functions the network-internal communication will improve.

And all that meets the patients full interest.

ONCONET: A Telematics Application Programme Projects Result

Teletreatment and Teleproctoring of a Home Self-adaptive Visual Field Treatment on PC/ TV Screen

BERND BLOBEL, PETER PHAROW, KJELD ENGEL

Institute for Biometrics and Medical Informatics, Otto-von-Guericke University Magdeburg, Leipziger Str. 44, D-39120 Magdeburg, Germany. E-Mail: Bernd.Blobel@MRZ.Uni-Magdeburg.DE, {Peter.Pharow,Kjeld.Engel}@Medizin.Uni-Magdeburg.DE

Key words: Health network; Security; Health Professional Card; Trusted Third Party services

Abstract: Resulting from projects launched by the European Commission within the fifth framework such as ISHTAR, TrustHealth, Euromed-ETS or MERDSEC, a security infrastructure has been implemented for enabling trustworthy communication and co-operation between healthcare applications. This security infrastructure consists of Health Professional Cards (HPC) as security tokens and a Trusted Third Party (TTP) providing the Public Key Infrastructure needed for the asymmetric kryptographic algorithms used for security services based on digital signatures such as strong authentication, integrity check, accountability but also for session key exchange. The solutions implemented are solely based on international standards. The generic security solution has been applied to establish a trustworthy healthcare network for seamless care of cancer patients. This shared care application enables the exchange of partial or complete Electronic Healthcare Records (EHCR) in oncology but also queries and statistical analysis about entities etc. The solution is used to support administration of cancer patients, clinical studies, education, research and development, but also quality assurance and optimisation of diagnosis and treatment.

1. INTRODUCTION

The shared care paradigm seems to be the only possible response to the challenge for mastering the crisis of health care systems by improvement of quality and efficiency of health systems in the industrial countries around the world. Shared care means specialisation and decentralisation accompanied by concerted communication and co-operation between the partners involved in the patient's care directly or indirectly. Therefore, communication and co-operation partners are the patient, health professionals (HP) including both health care professionals and administrative or technical staff as well as management in health care establishments, but also other personnel and institutions belonging to the health domain directly (facilities, laboratories, etc.) or indirectly (e.g., insurance companies, ministries). Even partners from outside the health domain must be considered, as vendors of systems, material suppliers, or libraries. Due to the different purpose of doing and the different distances to the patient and his/her care, these different communication and co-operation partners have different legitimation, rights and duties for using patient-related medical information. Considering these circumstances, communication and co-operation have to be realised in a trustworthy way bound to the specified purpose, requesting special environment reflecting the sensitivity of the data used.

The information exchanged and the services provided are sensitive in the sense of personal privacy of patient or staff involved as well as confidentiality of business information, etc. In that context, the sensitivity of personal medical information is caused by its legal, social, and psychological implications. Furthermore, information and services are exposed to different threats and risks. Therefore, communication and co-operation require appropriate security services to be implemented according to the policy agreed [1, 2]. Restricted to the concept of security, the paper describes security requirements and solutions in the context of the medical routine application ONCONET establishing the first secure German network of real medical applications.

2. THE REGIONAL CANCER REGISTRY MAGDEBURG/SAXONY-ANHALT

For the support of area-covering evenly high-degree care of cancer patients, so-called cancer centres or oncological centres with clinical cancer registries as documentation systems have been founded in Germany. With generous funding by the German Federal Ministry of Health, a network of meanwhile about 30 such institutions was installed also in Eastern Germany. Apart from clinical cancer registers, epidemiological cancer registers are also very helpful for the investigation of some cancer-related questions, as mentioned in paragraph 2. In 1993, the 5 "new" German federal states and Berlin have decided to continue the common epidemiological cancer registration in Berlin, called "Gemeinsames Krebsregister der Bundesländer Mecklemburg-Vorpommern, Brandenburg, Sachsen-Anhalt, der Freistaaten Sachsen und Thüringen sowie Berlins". This "Gemeinsames Krebsregister" both replaces and continues the "Krebsregister der DDR" which was the epidemiological cancer registry of the former GDR. It was a compulsory registration. The clinical cancer registers are the dominant registration sites for the data flow to this Berlin registry.

Within the cancer centres, which were founded

- to support the co-operation between the different institutions relating to labour-shared care of cancer patients,
- to improve quality and efficiency of cancer care,
- to promote research and development in oncology,
- to improve training and education,
- to elaborate standards for care and quality assurance etc.,

the clinical cancer registers should support the authorised user to achieve these objectives by available, integer, consistent and accountable information at the right time and at the right place. As the first one in Germany, the regional Clinical Cancer Register Magdeburg/Saxony-Anhalt has realised this concept including the health care establishments of the region and enabling secure direct access to the specialised co-operative Electronic Health Care Record (EHCR) for a catchment area with about 1.2 million inhabitants. So far, the shared care in oncology is being supported by a shared care IT solution.

Currently, the regional Clinical Cancer Registry Magdeburg/ Saxony-Anhalt comprises 57 clinics and more than 160 GPs specialised or at least commonly involved in the domain of oncology. The GPs are organised in the Oncological After Care Organisation of the Doctors' Statutory Association (Kassenärztliche Vereinigung) of the German federal state of Saxony-Anhalt (KV Sachsen-Anhalt). The communication has been increased from a bi-lateral communication between some partners and the registry to an extended communication and co-operation network called ONCONET. In the future structure, about 75 clinics and all the GPs interested will go online in the ONCONET.

The ONCONET's kernel consists of an ORACLE based registry application called „Giessener Tumordokumentationssystem“ (**GTDS**) which has been developed by the University of Giessen, Germany, using ORACLE tools as forms, reports etc.

In the cancer registry's initial phase, the applied security measures provided system authentication and confidential communication, using the MACS (Modem Access Control System) for analogue lines, the Kryptoguard LAN L3 boxes for ISDN lines between the secure server and secure external LAN (closed systems), and the ISDN Kryptoguard PC plug-in card for stand-alone (secure) PCs connected to the (secure) server by ISDN lines. Improvement and further development of the system is embedded in several security-related projects the Magdeburg Medical Informatics Department was involved in, as ISHTAR, TrustHealth, EUROMED-ETS, MEDSEC [2, 3, 4, 5]. For legal reasons, system-related security must be extended or even replaced by user-related security services. In that context, the European project "Trustworthy Health Telematics Applications" (TrustHealth-1) was analysing security requirements and specifying security solutions based on Health Professional Cards (HPC) and Trusted Third Party (TTP) services. As a result of that project, the Magdeburg Clinical Cancer Register became the first German healthcare application and the only TrustHealth-1 partner with an implemented advanced security mechanism ensuring strong authentication of users as well as integrity and confidentiality of data based on such standardised solutions. In the follow-up TrustHealth-2 project framework continuing the TrustHealth-1 project, large-scale demonstrators of really existing medical application should be implemented and evaluated using the European security infrastructure of HPC and TTP services

3. HEALTH PROFESSIONAL CARDS

Confirming and continuing "Trusted Health Information Systems" (THIS) project chaired by Gunnar Klein [6], smart cards have been defined as appropriate tokens securely bearing both the private keys needed for authentication, digital signature, and decryption (e.g. of session keys) and the certificates for those keys. Beside these identity-related (identity card) functionalities, some attributes concerning profession, professional qualifications and permissions to practice, but also other profession-related information (e.g. exposition of radiologists to radiation) are certified and stored in the card extending its character to a Health Professional Card (HPC). Contrary to the identity-related certificates, attribute certificates are not bound to keys. Generalising that principle, card functions may be separated according to certificates applied, as ID certificates, professional and organisational certificates, and application-specific certificates. More details on HPC are given in [7]. Suggesting open solutions, RSA has been favoured as the algorithm to be used. RSA, however, needs a crypto co-processor for processing the algorithm by that way increasing the costs for cards extremely. Therefore, alternatives are under development and exploration as, e.g., elliptic curves only using the standard smart card.

Meanwhile, the HPC has been standardised on the European level as CEN prENV 13729 "Health Informatics - Secure User Identification – Strong Authentication using Microprocessor Cards" [8] as well as consistently on the German national level as the HCP-Protocol [9].

4. TRUSTED THIRD PARTY SERVICES

Dealing with security services based on asymmetric cryptographic algorithms, a PKI based on a Trusted Third Party (TTP) structure is inevitable. A TTP summarises all the structural components which provide TTP services and belong to one or even several organisations. Beside the fundamental work of the European TrustHealth project, several other projects have been launched by the European Commission in the framework of "European Trust Services - ETS" over the last three years [10]. The original objective of ETS was the investigation and possible resolution of issues related to the creation of an appropriate enabling environment for the use and provision of security services related to principals or procedures such as authentication, non-repudiation, confidentiality and time stamping, by industry and commerce. These services may be offered by pan-European Trusted Third Party (TTP) Service infrastructures, as required by the market.

ETS has addressed the resolution of the issues and the measures necessary for the design, specification and market-driven implementation of a European Trusted Third Party Service infrastructure which will support the information security services needed to enable the European and global information infrastructure. Such a European information infrastructure for health has been piloted in series of EUROMED projects for creation of a pan-European Internet-based health network. As the German partner of the EUROMED-ETS project, such a pan-European Trusted Third Party (TTP) Service infrastructure has been established practically between Greece, Italy and Germany in 1997 including the universities of Athens, Calabria, and Magdeburg [4, 7, 11].

The goal of at all ETS has been to tackle, to the greatest possible extent, both the technical, economical, legal and regulatory aspects that govern the use of cryptography for authentication, confidentiality and non-repudiation, and to resolve the dilemma posed by the increasing importance of encryption in our information society.

Simply summarised, the paper-based world must be transferred properly into an electronic world. Considering identification, authentication, and certification of persons and their professional attributes, figure 1 describes formally the equivalencies to be provided.

The TTP generates and issues the cards, certifies the public keys, and combines them with valid timestamps. The certificates held in a public directory (directory service, revocation service) must be permanently available for enabling on the one hand signature verification (integrity service) as well as on the other hand the sending of encrypted data (confidentiality service). All user-related information concerning, e.g., the personal identity, professional qualifications and specialities are stored in certificates.

As mentioned above, the certificates may be hold in the card for off-line verification. This implies however possible disadvantages in the

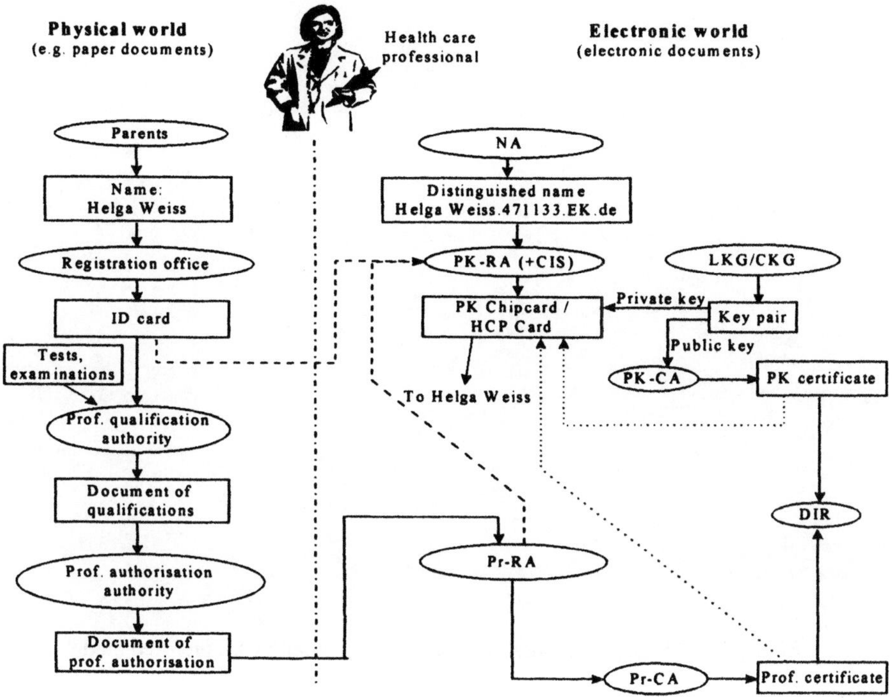

Figure 1: Electronic vs. Paper-Based World

context of updating and revocation of certificates in a given time interval.

To describe the structure of relevant Trusted Third Party services one must again emphasise that a TTP comprises all independent organisations which offer and are responsible for a defined TTP service. One girder of such an organisation should be a secure IT and communication system, which as a whole or in parts might be outsourced to another organisation. However, this is not the only or even the most important girder for a TTP to fulfil its basic objective: to offer security services with the necessary degree of (technical and business) functionality and assurance. Its formal or legal position within its service domain might be equally important.

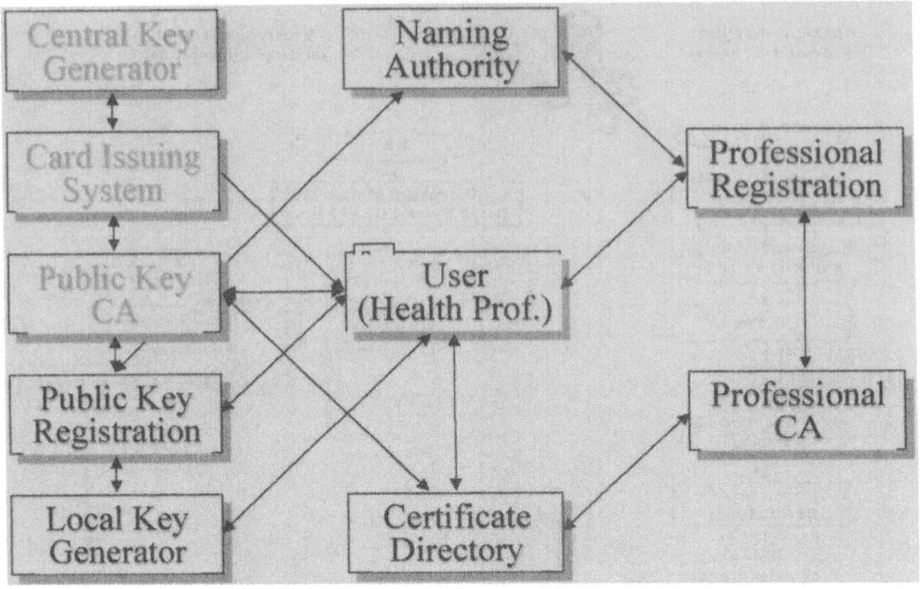

Figure 2: TTP roles and possible interactions – health care oriented model

Furthermore, a TTP service structure is not meaningful unless we define a set of roles and describe the objectives and tasks of these roles and the way how the roles interact. Figure 2 pictures the relevant roles and how the various roles might interact in a general TTP infrastructure.

Hereby a User is an individual or organisational entity. A Public Key Registration Authority (PK-RA) is an entity which uniquely identifies and registers users applying for the Digital Signature services provided, whereas a Professional Registration Authority (Pr-RA) is an entity which registers (and possibly authorises) individuals as health care professionals. The Naming Authority (NA) is an entity which appoints unique certificate names to users. The NA may also handle the naming of health care professional classes (e.g. physician), specialities (e.g., internal medicine) and possibly sub-specialities (e.g., nephrology). The Public Key Certification Authority (PK-CA) is an entity which certifies the linkage between the unique certificate name and the users' public signature or decryption key by issuing public key certificates digitally signed by the PK-CA. A PK-CA is also responsible for the revocation and re-issuing of public key certificates, whereas a Professional Certification Authority (Pr-CA) is an entity which certifies the linkage between the unique certificate name and the users' professional status by issuing professional certificates digitally signed by the Pr-CA. A Pr-CA is also responsible for the revocation and re-issuing of professional certificates. And last but not least the Card Issuing System (CIS) is an entity which issues chipcards containing (at least) the private keys of the users (card owners) for authentication, signature, and decryption. The generation of keys could be done by a Local / Central Key Generator (LKG/CKG) as an entity either located locally (by the user or PK-RA) or centrally (by the PK-CA or CIS) which generates the key pairs required. The certificates have to be stored in a Certificate Directory (DIR). It is an entity which provides the public key certificates, professional certificates, certificate revocation lists and possibly other information about users to other users at request.

In figure 3 the TTP roles and the interactions are shown which are specifically needed to influence functionality and security in the health care sector. These services should be established in health care institutions as the Physician Chambers (Ärztekammer) or the Doctors' Statutory Association (Kassenärztliche Vereinigung). Regarding the other roles presented in grey, the requirements are considered to be general requirements for overall confidence in TTP services provided in relation to the specified security policies and other relevant elements. More information might be found in [12].

As long as not all health professionals are organised in health professional chambers, a TTP structure for health must be established fulfilling both the legal framework and common requirements. Therefore, some authorities needed for provision of certain services must be implemented separately in the cancer centre framework. For protection of investment in technology and organisation, as much authorities as possible have already been installed within the future chamber structure.

5. THE SAXONY-ANHALT ONCONET

The ONCONET enables secure communication between the regionally distributed members of the Cancer Center Magdeburg/Saxony-Anhalt as well as between them and the Clinical Cancer Registry as an EHCR system for oncology. Additionally to the communicating systems, an infrastructure for trustworthy communication has been installed consisting of a communication server, a proxy server for functional interoperability, and appropriate TTP services. As an interim solution, the CA Management System of the SecuDE GmbH Darmstadt, developer of the SecuDE™ (Security Development Environment) products used in our environment, has been applied [13]. Figure 3 shows the principle architecture of the ONCONET solution.

For the analysis of system requirements, system design, specification, implementation, and maintenance a generic component model approach and the UML methodology has been applied [14, 15, 16] which is also usable to describe different views from legal issues up to the technical implementation details according to [17].

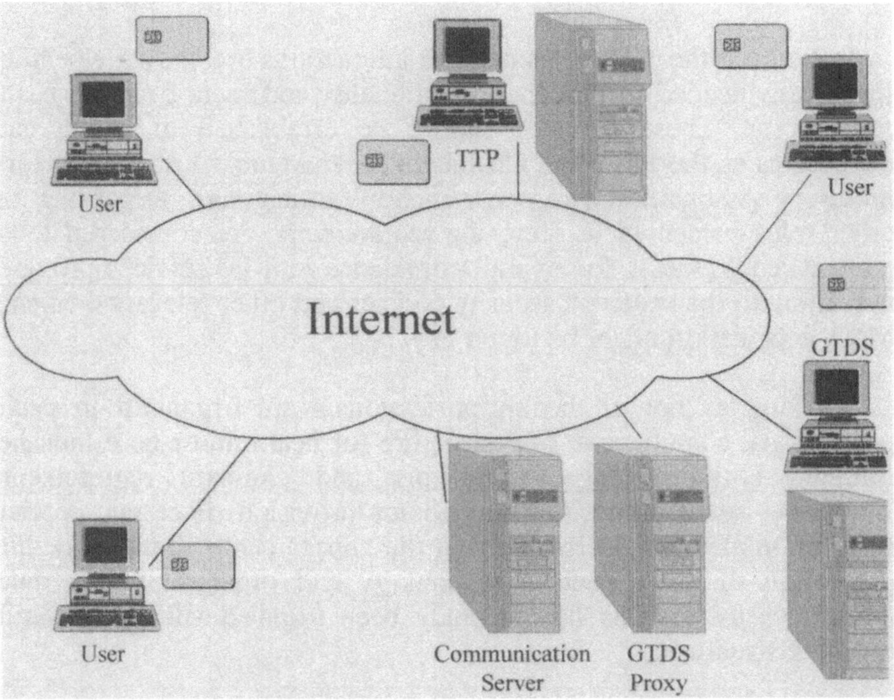

Figure 3: Saxony-Anhalt ONCONET Schema

One medical scenario deals with the secure asynchronous communication between a health professional (i.e. by law the responsible doctor) and the Medical Documentation Assistants at the registry site for transferring the doctor's report. Usually, the doctor's report will be created using text processor systems as MS Word. The trustworthy communication is only provided between authorised users, guaranteeing confidentiality and integrity of information transferred. Additionally, the user's role as doctor is verified by checking the corresponding attribute certificates. The integration of signature and encryption into the MS Word application is provided via DLL by a Visual Basic for Application (VBA) macro programmed by the Magdeburg Medical Informatics Department. The solution is based on both the HPC and the SECUDE development system. The strong mutual 3-way authentication deployed corresponds to the ISO specification [18, 19]. The secure information transfer is realised via the Secure File Transfer Protocol (SFTP) meanwhile approved as HL7/ANSI Standard. This universal protocol, which enables the communication of any type of data with unlimited extension, is widely used in many contexts (e.g. also HL7, images) around the world. It is described in more detail in [20, 21, 22, 23]. For legal reasons, all incoming information signed by its originator is securely stored in a separate electronic archive at the register site. Figure 4 shows the sequence diagram giving a detailed description of the procedures of doctor's report transfer.

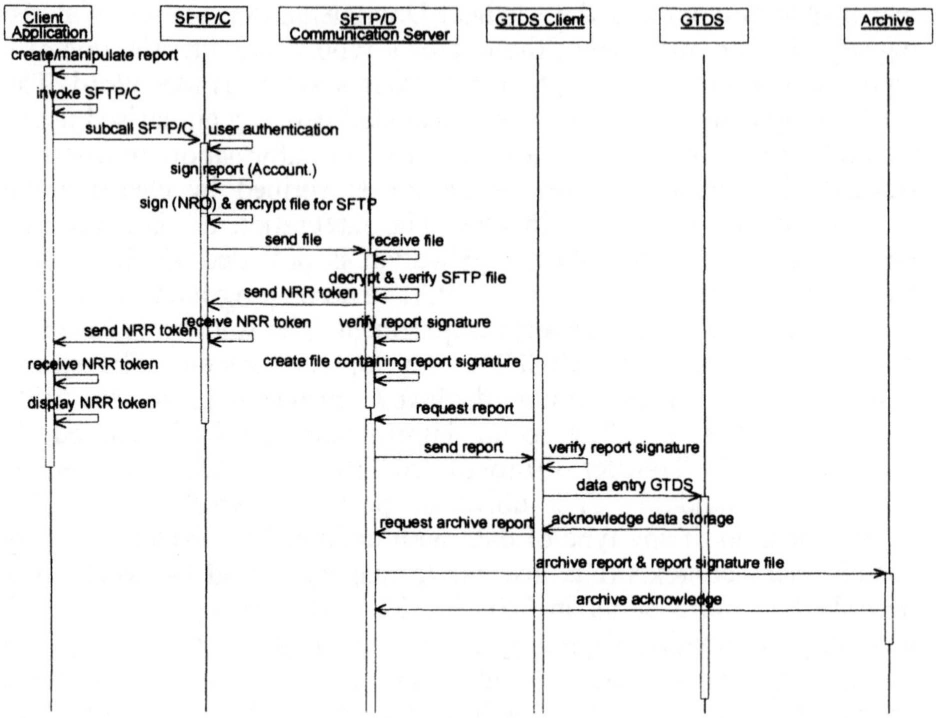

Figure 4: Sequence Diagram of the Doctor's Report Transfer Use Case

Another ONCONET scenario deals with general communication between the health care establishments involved in the network partners as, e.g., GP-GP or GP-clinic communication. In that context, doctor's report for referrals, discharges, or collegial information, but also agreements or schedules in reaction of planned activities of the co-caring institutions accessable to the partners in real-time may be transferred. Regarding his/her patients, the corresponding information can be retrieved by the doctor using SQL queries to the oncological EHCR system established at the register site. This third ONCONET scenario enables the doctor to transfer the whole health record of one of his/her patients. Using predefined or even free SQL queries, he/she can get aggregated data of his/her patients. Thus, "statistical" overviews related to the organisation, to certain entities characterised by specific diagnoses or therapies as well as other care-related information usable, e.g., for process optimisation, research, education, and quality assurance can be obtained. Figure 5 demonstrates the schema for the request and provision of data from the registry. The solution for trustworthy communication between the strongly authenticated principals doctor and record system is the same as described above. For application security reasons, there is no direct access of external users to the registry system. This is true also in the case of the third scenario which does not include any human activity at the EHCR site. A proxy retrieves the information on behalf of the user presenting his/her role to the system for the right authorisation. On behalf of the register, the data permitted are securely transferred to the requestor by that proxy signed and encrypted using the requestor's public key. No unauthorised principal can decrypt the data, if listening to the communication line.

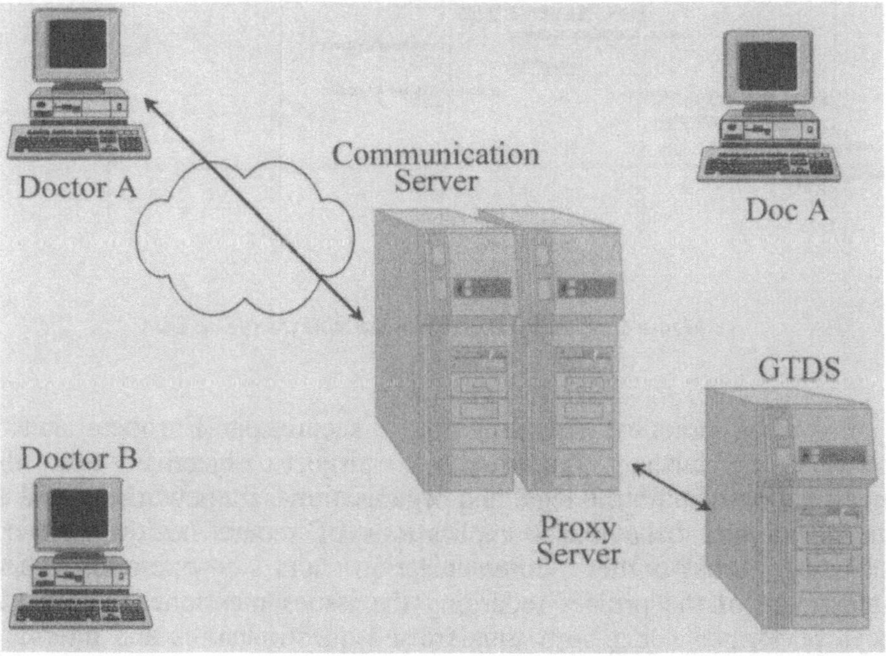

Figure 5: Schema of Patient Information Request

Figure 6 presents the sequence diagram of the request of patient information by SQL queries.

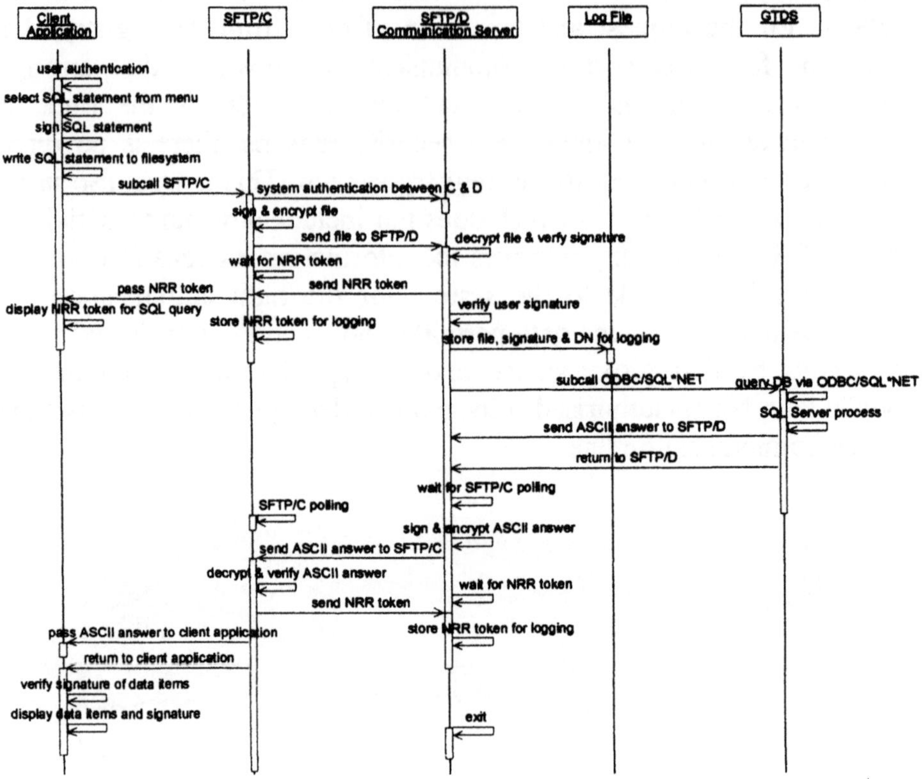

Figure 6: Sequence Diagram of the SQL Query Use Case

Providing practical solutions for a secure pan-European health network, a further TrustHealth–2 project objective was the investigation of both the legal and organisational framework as well as the social and behavioural implications of secure health networks including cross-border communication and co-operation. The assessment of the project regarding the issues mentioned before has been performed using both standardised questionnaires and interview techniques [3, 24]. The evaluation procedure is not finalised yet. For the oral presentation of the paper however, the results will be available.

CONCLUSIONS

The presented solution is one of the European TrustHealth project's large scale pilots, but also the first demonstrator of the German HPC.

Based on the results of several European projects and international (especially American) standardisation initiatives on health informatics the Magdeburg Medical Informatics Department is involved in, a trustworthy health network has been specified, developed, and implemented. If other German health network initiatives usually support not health-related but common services as secure e-mailing, the ONCONET aims communication and co-operation between medical applications. Therefore, medical procedures are facilitated such as planning and organisation of diagnosis and therapy (e.g., chemotherapy based on a huge set of therapy schemas), design and application of clinical guidelines, but also administrative actions like appointments, scheduling, or reports. Furthermore, research and development issues are supported directly. In that context, clinical and epidemiological studies, and quality assurance, but also optimisation of diagnosis and therapy should be mentioned. Finally, education can be enabled as well as outlines, trends, and decisions can be derived for health policy and administration based on the information available.

Beside the cancer care related services mentioned above, also common services for health are provided such as doctor-doctor communications or the transfer of CT image from external institutions to the University Hospital Magdeburg to ask for and to get second opinions.

The Saxony-Anhalt ONCONET has been designed for improving the care of cancer patients. However, the network could be expanded as a general health network of the region success and acceptance assumed.

The security infrastructure specified, developed, and implemented is based on international, European, and German standards and specifications. This concerns the standardised HPC applied for identification, authentication, digital signature, decryption. Additionally, it provides special functionalities as role assignment, registration of education on the one hand or exposition to particular conditions (e.g., radiation) on the other one. But also appropriate TTP services (public key infrastructure) following standards have been implemented. Regarding the communication protocols, open solutions and widely used standards, such as secure FTP or mailing procedures have been applied. On the cancer registry site, a well specified schema for authorisation and access control has been implemented for guaranteeing a trustworthy and legislation-conformant EHCR environment.

ACKNOWLEDGEMENT

The authors are indebted to the European and the German TrustHealth-2 project partners. Especially, they'd like to thank the colleagues of the Physician Chamber Saxony-Anhalt as well as of the health care establishments involved in the ONCONET for their engagement.

REFERENCES

1. Blobel B, Katsikas SK (1998) Patient data and the Internet - security issues. Chairpersons' introduction. International Journal of Medical Informatics **49,** pp. S5-S8
2. The ISHTAR Consortium. http://www.ehto.be/projects/ishtar
3. The TrustHealth Consortium. http://www.ehto.be/projects/trusthealth
4. The EUROMED-ETS Consortium. http://euromed.ece.ntua.gr
5. The MEDSEC Consortium. http://www.etho.be/projects/medsec
6. The THIS Consortium. http://www.etho.be/projects/this
7. Blobel B, Pharow P (1997) Security Infrastructure of an Oncological Network Using Health Professional Cards. In: Broek L van den, Sikkel AJ (eds) Health Cards '97, Series in Health Technology and Informatics Vol. 49, IOS Press Amsterdam, pp. 323-334
8. CEN TC 251 (1999) prENV 13729: Health Informatics - Secure User Identification – Strong Authentication using Microprocessor Cards (SEC-ID/CARDS)
9. HPC (1999) The German HPC Specification for an electronic doctor's licence. Version 0.81, February 1999. http://www.hpc-protocol.de
10. The ETS Projects. http://www.cordis.lu/infosec/src/ets.htm

11. Katsikas SK, Spinellis DD, Iliadis J, Blobel B (1998) Using Trusted Third Parties for secure telemedical applications over the WWW: The EUROMED-ETS approach. International Journal of Medical Informatics **49**, pp. 59-68

12. Pharow P, Blobel B (1999) Trusted Third Party Services for Internet Security. In: Mastorakis NE (edr) Recent Advances in Signal Processing and Communications, World Scientific and Engineering Society Press, pp 379-385

13. SECUDE (1999) A General Purpose Security Toolkit. Specification of the SECUDE Software. http://www.darmstadt.gmd.de/secude

14. Blobel B, Pharow P, Roger-France F (1999) Security Analysis and Design Based on a General Conceptual Security Model and UML. In: Sloot P, Bubak M, Hoekstra A, Hertzberger B (eds) High Performance Computing and Networking, Lecture Notes in Computer Sciences 1593. Springer, Berlin, Heidelberg, New York, pp 919-930

15. Blobel B, Roger-France F (2000) A Systematic Approach for Secure Health Information Systems. International Journal of Medical Informatics (submitted)

16. Eriksson H-E, Penker M (1998) UML Toolkit. John Wiley & Sons, Inc., New York

17. Blobel B (2000) Application of the Component Paradigm for Analysis and Design of Advanced Health System Architectures. International Journal of Medical Informatics (submitted).

18. International Organisation for Standardisation: Information technology, Open Systems Interconnection, Security frameworks for open systems, multiple Parts (1-7).

19. International Organisation for Standardisation: Information technology, Security techniques, Entity authentication, multiple Parts (1-5).

20. Blobel B, Pharow P, Engel K, Spiegel V, Krohn R (1999) Communication Security in Open Health Care Networks. In: Kokol P, Zupan B, Stare J, Premik M, Engelbrecht R (eds) Medical Informatics Europe '99, Series in Health Technology and Informatics Vol. 68. IOS Press, Amsterdam, pp 291-296

21. Blobel B, Spiegel V, Krohn R, Pharow P, Engel K (1998) Standard Guide for EDI (HL7) Communication Security, http://www.hl7.org

22. Blobel B, Spiegel V, Krohn R, Pharow P, Engel K (1998) Standard Guide for Implementing EDI (HL7) Communication Security, http://www.hl7.org

23. CEN TC 251 (1999) prENV 13608: Health Informatics - Security for Healthcare Communications,

24. Wenzlaff P, Blobel B, Pharow P (1999) Health Professional Cards: Awareness for Security in Health Care Networks. In: Sicurello F (edr) Health Cards '99, XASI, Milan, pp 121-125

Integration Of TLS In An Internet/Intranet-based Application

LUTZ VORWERK, F. LOSEMANN, CH. MEINEL
Institut für Telematik e.V, Bahnhofstrasse 30-32, D-54292 Trier, Germany

Keywords: telemedicine, radiology, multimedia, security, transport layer security, Java

Abstract: In telemedicine a growing need for security is obvious. There are several approaches to integrate security in HIPAA (Health Insurance Portability and Accountability Act) or standards like DICOM (Digital Imaging and Communication in Medicine). These definitions are independent from the kind of implementation or use of programming languages. Here an implementation will be described which integrates a secure transfer of data in an existing system by using the transport layer security (TLS) and Java.

1. INTRODUCTION

This paper describes how to integrate encryption in the communication process of a radiological department with local physicians. As a helping tool in the development serves the Java security extension[8] and the Java Cryptix extension[7]. An example is the workflow of the creation of reports in radiology. Let us have a closer look at this workflow.

2. IMAGING AND IMAGE MANAGEMENT

The field of application radiology has different techniques. The choice of the appropriate method should depend on keeping the hassle for the patient with x-rays as low as possible. The approaches used today are improved with regard to the emission of x-rays, but when using techniques like Computer Tomography (CT), which produce a series of images, one ought to think about how often this technique should be applied to a patient. The CT uses x-rays to produce cross-section images of the patient being examined. The technique to use magnetic fields instead of x-rays is called Magnet Resonance Tomography (MRT). This technique produces images from the resonance of hydrogen atoms evoked by magnetic fields. It is less dangerous with regard to x-ray burden, but expensive and dangerous if pieces of metal are contained in the body of the patient. If these pieces are not discovered in time and the technique is applied to the patient, these pieces will be influenced by the magnetic field and can be responsible for internal injuries. Another technique is the use of ultrasonic waves to produce an image. This technique is appropriate for the pre-examinations or examinations which do not need high-resolution capture systems. The techniques described here can be combined with the use of a contrast medium which is also inconvenient for the patient. It is obvious that with respect to the patient, the techniques and the use of a contrast medium should only be applied if necessary. Redundancy can be prevented if the digital images of a patient are archived in an image archive, which is central or de-central. If a physician ordered an radiological examination, the order and the digital images are stored together in a database. If the patient changes his physician, the images are available for the other physician. This method saves a lot of costs. Several systems named HIS (Hospital Information system) /RIS (Radiology Information System) in case of management of patient data others named PACS (Picture Archiving Systems) and are used for image management. There are approaches to combine a HIS/RIS with a PACS. Such systems are known as HIPACS. Some hospitals uses DICOM imaging devices and try to integrate it in the workflow.

3. SITUATION IN A RADIOLOGICAL DEPARTMENT

Assume the working conditions in a radiological department with single place systems without a connection to the internet: One important feature of an application used in this environment is that this application has an easy and fast operating user interface which is easy to understand. This is because findings have to be performed quick and under time pressure.

To realize a system which finds the approval of most physicians, it has to be developed in JAVA. More precisely, JAVA provides the functionality to implement a user interface mentioned above. This programming language allows to connect to the order database where the physician can deposit his or her order. The order is then processed by the radiological department, which orders, in turn, existing images of the patient and decides if there are available images and if these images are sufficient to answer the requested order of the physician. If yes, the radiologists perform the findings by using these images and send the report to the physician. If not, they tell the patient to make an appointment to produce new images. This procedure serves the physical health of the patient.

Among the existing components of this system is a multimedia editor[4,5,9(Client)] which will be connected via sockets to a server that manages the data to be transferred. A socket is one end-point of a two-way communication link between two programs running on the network. These sockets provide a secure protocol mechanism capable to store the reports and orders done by physicians in a secure way.

3.1 Integration of Encryption

The new system component presented here uses the javax - API (application programming interface) to implement a secure transfer between two locations[8]. One location has an Applet (i. e. the multimedia-editor or a text editor) which uses the implemented TLS (transport layer security)[1] to communicate with a Servlet which works on a database to store incoming requests and process them. Incoming requests are URLs that refer to objects to be processed. Those URLs contain keys to decrypt the data of the referenced object.

The first thing you have to consider here is the definition of the TLS protocol definition. After that, an adaptation of the TLS protocol to the structure of the hierarchical ordered security classes from the javax-API is necessary. To perform the adaptation we have restricted the used encryption algorithms to those which are actually considered as highly secure. The management of used algorithms will be performed by a database which stores the compiled Java classes. The database will be used by a Java Servlet. The communication between the database and the Servlet is handled in a secure way to prevent interior attacks. Information send to the database will be encrypted by PGP (Pretty Good Privacy). The Servlet governs a key table stored as an pgp-encrypted object in the database. During initialization the Servlet asks the user for login and password in order to be capable to decrypt the key table. If the key table is available the Servlet is able to read and store objects in the database. New keys will be created for new objects. We are using the Cryptix Java Security extension as a helping tool. It provides the majority of the encryption algorithms and is a valuable basis for our development. How to use TLS within the system is shown in Figure 1.

Cryptix supports encryption algorithms to perform symmetric encryption, asymmetric encryption, block ciphers, and digital signatures. The following list contains the encryption algorithms we use to implement our TLS-Protocol.

3.2 Symmetric Encryption

Symmetric encryption use the same key for encryption and decryption. DES-EDE3 is chosen for symmetric encryption because it is considered as secure in accordance with an asymmetric encryption: This is 3 key-Triple DES. The conformance of this algorithm used in Cryptix have to be checked, yet. The data, checked against SSLeay only and therefore needs further comparisons.

3.3 Asymmetric encryption

Asymmetric encryption use a key pair consisting of a public and private key. In case of encryption the public key is used for encryption of data. The private Key is used for decryption. The asymmetric encryption method 'RSA' (Rivest, Shamir, Adleman) has been patented in the United States but not in Europe. The patent expires on 20-th September 2000. Within Europe, differing methods are used[3,6]. Key encoding allows to generate RSA key pairs by using a modulus of less than 768 Bit. RSA will only be useful in the future if it is possible to perform the modulus of a length of 1024 Bit or more. RSA is used within the protocol-implementation to guarantee compatibility.

3.4 Cipher Modes

CBC (cipher block chaining) is appropriate because it allows block ciphering for predictable connections[2]. There are other block ciphers but they are not convenient enough. They are either insecure and require additional treatment to guarantee security or they have features not needed for the implementation of TLS. A block cipher is not yet included in the application. Instead of this the data to be transferred is divided into 512 Kbyte blocks. These blocks will be then encrypted by using the symmetric encryption method DES.

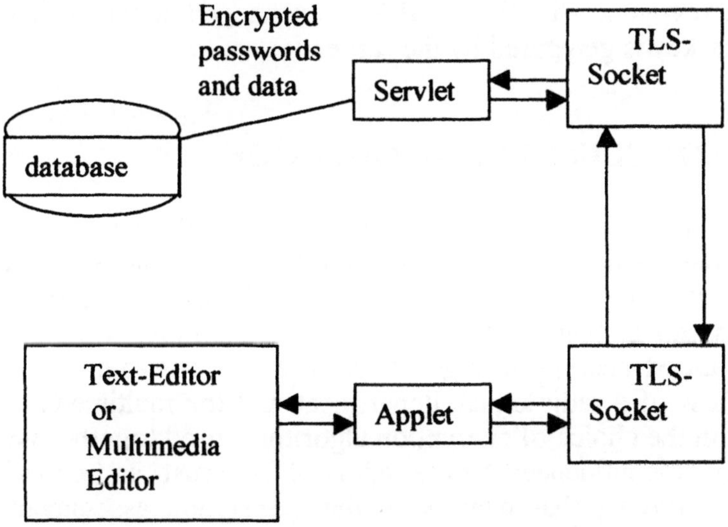

Figure 1. Integrating a TLS connection

3.5 Signature Modes and Message Digests

Digital signatures are used to identify the responsibility of a digital document. The document is a certificate signed by the provider of this certificate, a text or other digital data. As a hash function for signatures it is most common to use MD2 (MD = message digest) or MD5. We cannot recommend MD2, MD4 and MD5 though since they have been proven insecure. Appropriate hash functions are SHA-1 (Secure Hash

Algorithm 1) or RIPEMD-160 (RACE integrity primitives evaluation). We choose SHA-1 for our implementation.

4. BREAKING DOWN THE TLS DEFINITION FOR OUR PURPOSES

This section describes how the TLS implementation works. First the Server sends the public key to the client. This key is included in a certificate. Next, the client sends the public key (RSA key generation) to the server. This key is included in a certificate as a response to the request of the server. The client and the server both verify the certificate. After positive verification the key can be used to exchange messages. First, the symmetric key algorithm is negotiated. A symmetric key, generated from 3DES, is exchanged between client and server. This key is generated by the server.

5. CONCLUSION AND OUTLOOK

The system presented here is capable to use TLS for the transfer of individual data which is regarded as confidential and therefore have to be encrypted. The transport layer security is chosen to guarantee high security and for testing out this protocol with the most popular and as secure regarded encryption algorithms. The test environment will be the workflow of a radiological department and the multimedia editor. Influence on the choice of encryption algorithms used together with the TLS protocol is influenced by the report of Teletrust[3] and experiences gained in a project that deals with the description assessment of a secure infra structure[3].

REFERENCES

1. Transport Layer Security Working Group, Dierk, T., Allen, Ch., 1997, The TLS-Protocol Version 1.0.

2. Alfred J.Menezes, van Oorschot, P. C., Vanstone, S. A., 1996, Handbook of Applied Cryptography.

3. TeleTrust Deutschland e.V., 1998, Kryptographische Verfahren im Gesundheits- und Sozialwesen.

4. Vorwerk, L., Meinel, Ch., 2000, A multimedia editor for making findings in radiology. In *CBMS (Computer based medical systems).*

5. von Gehlen, S., Vorwerk, L., Eichelberg, M., Jensch, P., 1999, A multimedia editor for radiological reports based on the DICOM Supplement Structured Reporting. *In CARS `99, Computer Assisted Radiology and Surgery,*(H.U. Lemke, M.W. Vannier, K. Inamura and A.G. Farman), 438-442, Elsevier, Paris.

6. Vorwerk, L., Lüpken, B., Radke, C., Engel, T., Meinel, Ch., 2000, Bericht zur Bestandsaufnahme zur Sicherheitsinfrastruktur in DIZ. Institut für Telematik.

7. Systemics Ltd on behalf of the Cryptix Development Team, 1997, Cryptix is a trademark of Systemics Ltd., (http://cryptix.moremagic.com/).

8. Sun Microsystems, 2000, JavaTM Secure Socket Extension 1.0.1, Inc. 901 San Antonio Road, Palo Alto, California, 94303, U.S.A.

9. Vorwerk, L., Entwicklung eines Multimedia-Editors für die Befunderhebung in der Radiologie auf der Basis des DICOM Supplement Structured Reporting, (Master Thesis), Carl von Ossietzky Universität Oldenburg, November 1998.

Virtual Polyclinic

A Web-based system for medical teleconsultation

[1]SVEN LONCARIC, [1]EDGAR PEK, [1]MIROSLAV SILOVIC AND [2]ANAMARIJA MARGAN

[1]*Faculty of Electrical Engineering and Computing, University od Zagreb, Unska 3, 10000 Zagreb, Croatia, email: sven.loncaric@fer.hr, Fax: 385 (1) 612-9652* [2]*Practise for Internal Medicine, Cres, Croatia, email: anamarija.margan@ri.tel.hr*

Key words: telemedicine, medical teleconsultation, island telemedicine, electronic health record, Internet-based health record

Abstract: East Adriatic coast is characterized by a large number of islands that have always presented a challenging task in implementation of efficient health care in this part of Croatia. One of the most difficult problems is absence of medical specialists because specialist practices on islands are not financially self-sustainable. In particular, successful implementation of health care on the North-Adriatic islands of Cres, Mali Losinj, and neighboring islands has shown to difficult, especially during the high tourist season when the island population becomes much larger then during the out-of-season period. In this paper, we present a web-based system for medical teleconsultation support on islands and other rural areas. The system provides general practitioners on islands with tools to communicate patient information to specialists, typically located in large mainland medical centers. The system supports transport of multimedia medical information. Specialists use the system to transmit medical consultation back to the referring physician. Special attention is paid to security aspects.

1. INTRODUCTION

Providing health care services in rural areas has shown to be a challenging task. Low concentration of population makes health care instutitions not self-sustainable so typically only small ambulances with general practitioners are available in rural areas. In such situation, one of the biggest problems is lack of medical specialists. Patients requiring specialist opinion must travel to distant medical centers in big towns which is often not feasible due to the patient state or in medical emergencies. Such state of affairs can be dramatically improved by use of electronic means of communication. Rapid Internet development in recent years has made possible new medical applications including telemedicine [1,2]. Telemedicine helps connect rural areas to large medical centers through exchange of medical information between distant locations.

In this paper, we describe a web-based system for medical teleconsultation between small island medical centers and large medical centers located in towns. The system is currently under development. In Section 2, we describe the concept of the system. Details of database implementation are presented in Section 3. User interface and security aspects are discussed in Sections 4 and 5. Conclusion is given in Section 6.

2. VIRTUAL POLYCLINIC SYSTEM CONCEPT

2.1 Motivation

Lack of medical specialists in rural areas such as islands is a motivation for distant consultation of specialists in large medical centers. Croatian coast of Adriatic sea is characterized by a large number of islands many of which have population permanently living there. During the high tourist season there is a large increase of population also requiring medical services. General practitioners and internists require opinion of specialists in order to provide higher quality health care. Conventional procedure requires that the patient travels to a distant medical center. This is often impossible due to a high patient risk, or weather and sea conditions. When transport is possible it represents additional expense that has to be refunded through the health insurance system. If we were able to eliminate need for travel, it would be very beneficial medically and financially.

In this paper we describe Virtual Polyclinic (VP), a web-based system that integrates a multimedia teleconsultation system and an Internet-based health record system. VP helps eliminate the need for travelling between the Croatian islands and large medical centers. VP allows physicians to request and obtain medical consultation from specialists over the WWW.

Security is a very important aspect in many applications of Internet today such as e-commerce and e-medicine. VP employs many techniques for protecting patient privacy. These techniques include data encryption, user authentication, audit procedures, and database integrity protection. The developed system provides a high level of security.

2.2 Virtual Polyclinic System Architecture

VP system consists of a web server and a database server. Web server provides interface to users who access the system over the web. A database is used to store all information in the system. The system is implemented on Linux platform.

The web server is Apache [5], extended with mod_python for dynamic content, and with mod_ssl, which interfaces Apache with OpenSSL library, for secure authentication and data encryption. Webpage formatting is handled by XSLT (XSL Transformations) [6]. Mod_python transforms the output from SQL queries into raw XML trees, which are transformed into XHTML by Sablotron XSLT processor using the stylesheets. This way, web page design is separated from web page content. PostgreSQL [7] is used as a database server. It is interfaced to Python via the built-in python module. The architecture of the VP system is shown in Figure 1.

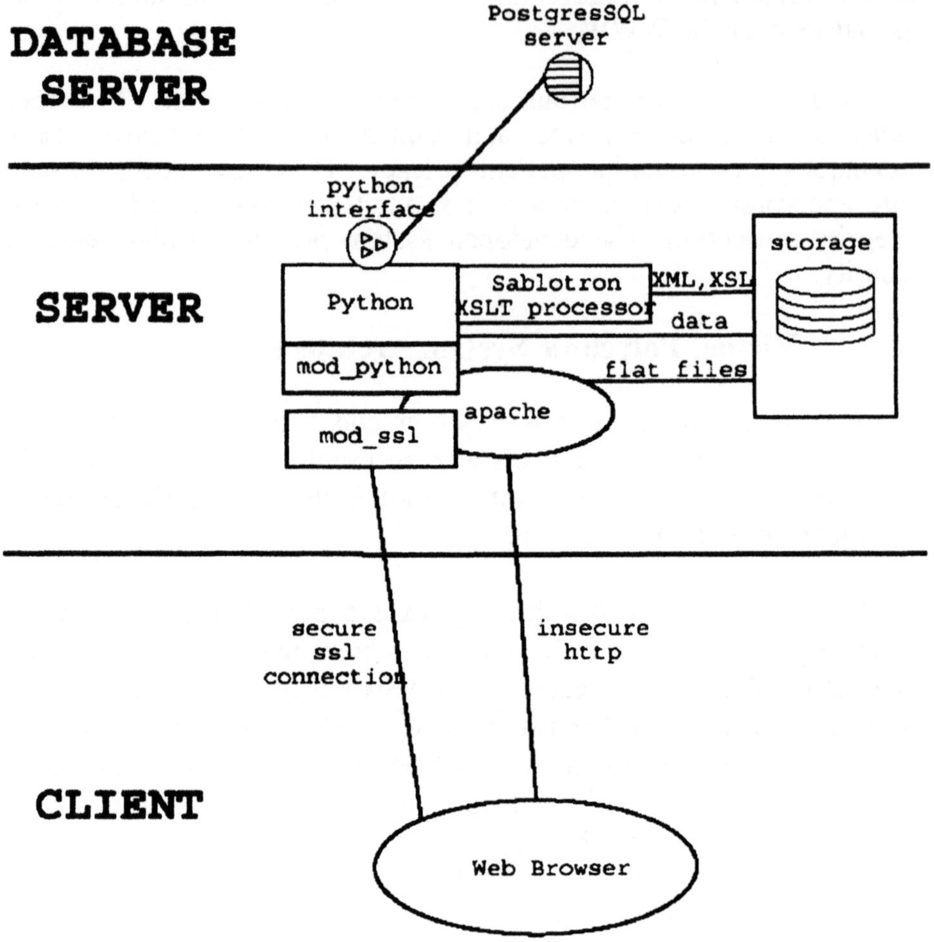

Figure 1. Block diagram of the Virtual Polyclinic system architecture.

3. DATABASE DESCRIPTION

In this section we describe the structure of the database, which is used in VP system. The database can be divided in two main parts. The first part stores teleconsultation data while the second part contains electronic health record data [3]. The database contains:

- User (physician) data, which contains user ID, personal information, medical specialization, login, and password

- Audit information about user access to the system (user identity, actions taken)
- Teleconsultation requests which are being sent by physicians
- Patient records which consist of: personal information, immunizations, medications, allergies, laboratory results (text, images), problem list, and hospital visits

The patient information is recorded through time, so a common attribute of all mentioned data is date and time. Entity-relationship (E-R) model [4] is shown in Figure 2. Note that most of the relationships are binary (only one is reflexive).

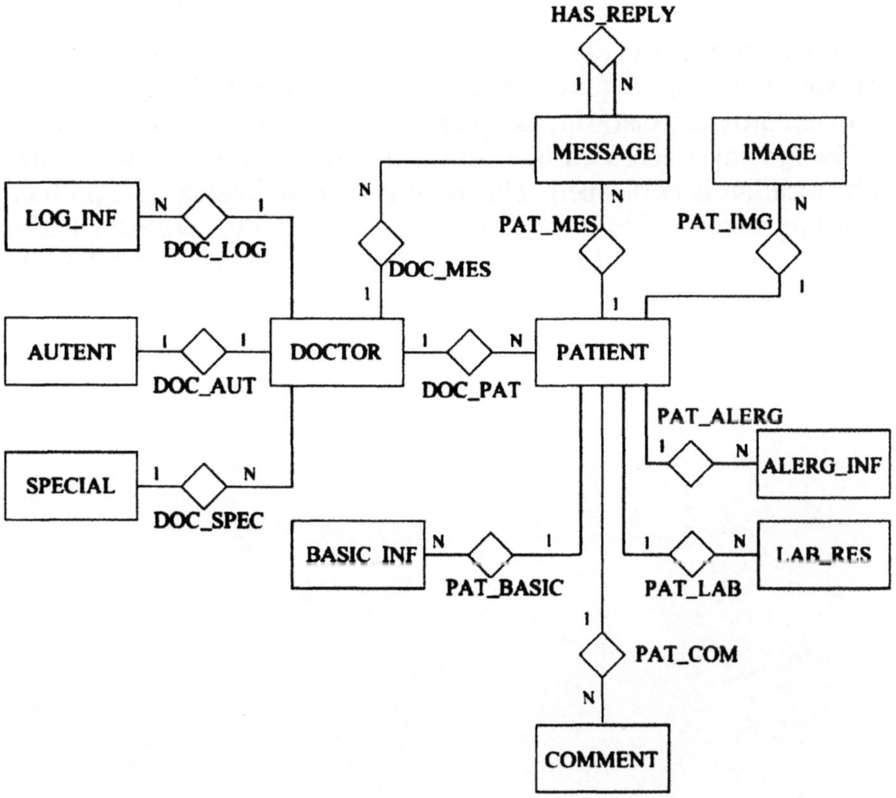

Figure 2. Entity-relationship (E-R) model for VP database.

E-R diagram shows entities and how entities are mutually related, e.g. entity *SPECIAL* and entity *DOCTOR* (stands for specialization) are related through relationship *DOC_SPEC*; mapping between entities is 1:*N*, which means that a physician is specialist in only one area, but several physicians can specialize in a single area. In the same manner we can describe most of the other relations between entities.

One special relation mentioned above is reflexive relationship, in our E-R diagram reflexive relationship is *HAS_REPLY*, and it is connected with entity *MESSAGE*. In this case reflexive relationship is used to show that entity *MESSAGE* can have replies, replies have same attributes as *MESSAGE* so entity *MESSAGE* have two roles, the first as message and the second as reply.

4. WEB-BASED USER INTERFACE

User interface for VP modules is based on WWW technology. In this way, user only requires Internet browser to use the system. This approach provides platform independence of the developed system.

The system is accessed by opening the initial web page where user authentication is performed. This page is accessed using SSL protocol to ensure secrecy of exchanged information. The user provides his/her username and password to access the following pages. The initial web page is shown in Figure 3.

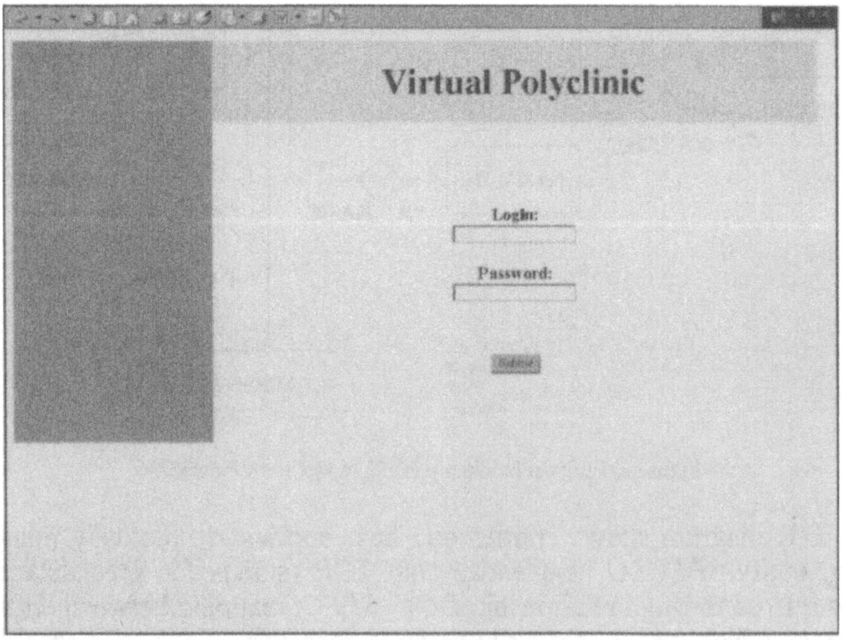

Figure 3. The authentication web page.

4.1 Teleconsultation Module

Teleconsultation module provides users with functionality required to conduct remote teleconsultations. A teleconsultation request contains information about the patient which is located in electronic patient health record and a note from referring physician describing the request.

To achieve minimal response time the physician is also notified of a pending consultation request by means of GSM SMS (short message service). In this way, the physician receives notification about the request on his mobile phone.

User interface of the teleconsultation module consists of a set of web pages visited by the user. The initial page is shown in Figure 4.

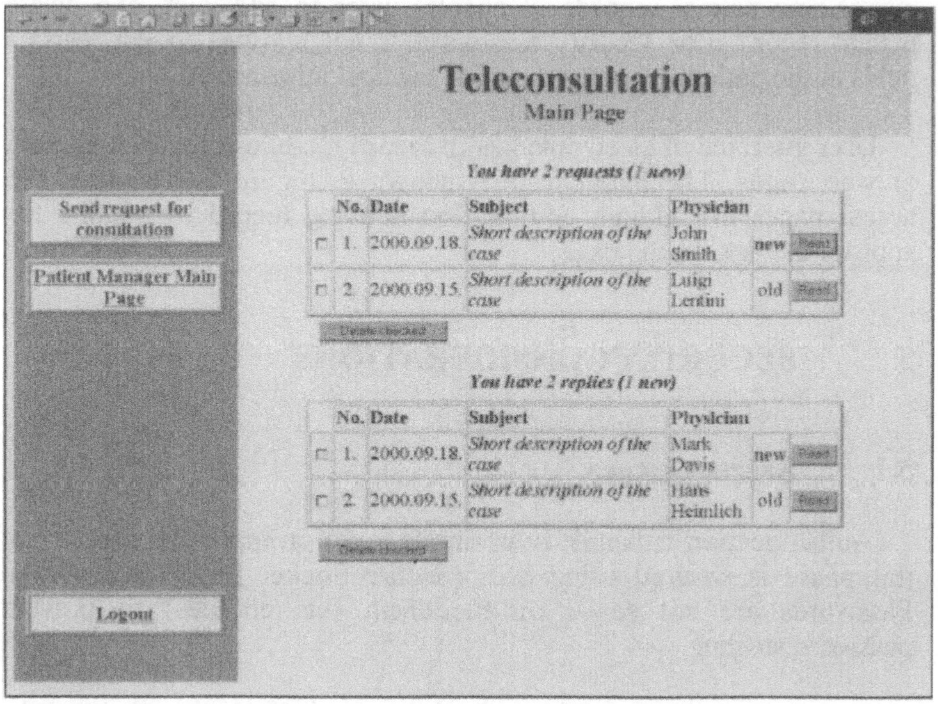

Figure 4. Teleconsultation module main page.

4.2 Electronic Health Record Module

Patient record as a data structure that helps to describe important facts considering patient's health has been used since the fifth century

B.C. when Hippocrates proposed purpose of the patient record. His observations were noted in chronological order, so this type of record is called time-oriented medical record. Improvement of the medical record organization lead to source-oriented and problem-oriented medical record (for details see [3]), where data within sections are in chronological order.

Our implementation of an electronic patient record is based on the source-oriented concept. According to that concept the record content is structured according to the method by which it was acquired, e.g. laboratory results, allergy information, etc. Data within each source is organized in chronological order.

VP electronic health record module provides functionality for maintaining patient records. It enables users to add, edit, and delete patient record. For security reasons it is not allowed to edit certain fields in the patient record containing medical information.

User interface of electronic health record module is realized as a set of web pages. The initial page is shown in Figure 5. The user can access pages for adding, editing, and deleting record by visiting the appropriate link on the page.

5. SECURITY CONSIDERATIONS

5.1 Secrecy issues

Authentication is handled via simple login prompt. The security of this phase is ensured using SSL (Secure Socket Layer) encryption. Passwords are not saved on the client (via cookies) to prevent password stealing.

In the current system, the authority space is 'flat' - in other words, each user has the same level of access. This may change, depending on the requirements. Database itself controls the privileges. System administrator is required to have physical access to the database server.

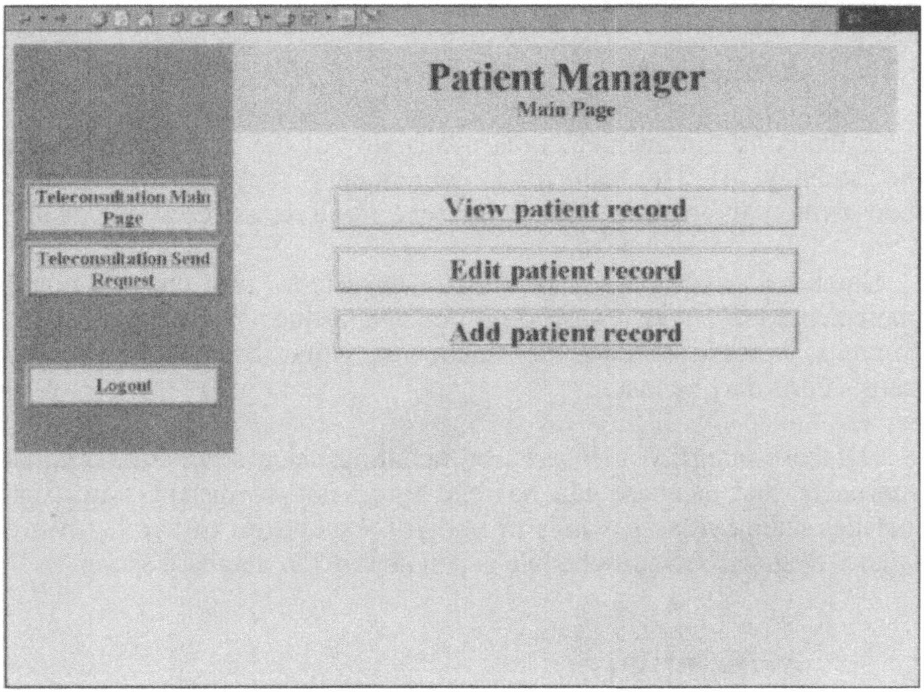

Figure 5. Electronic health record main page.

Accountability aspect is handeled in the following way. Database uses stored procedures to log all the transactions (and all the database access goes via stored procedures). The webserver does not allow access that does not get logged (still, gaining root access on the webserver may allow the cracker to attack the database itself).

Database has integrity checks implemented, up to the limitations of SQL. This includes referential integrity checks. SSL protects data integrity in transit (including man-in-the-middle attacks).

To protect confidentiality the data is only served to authorised users, using the authorization scheme described above. SSL is used to encrypt the server responses in transit.

To ensure database protection we have used the following methods. Database is on a separate machine, which does not have direct network access (it is firewalled and limitted to the access from the webserver itself). Gaining the root access on the server may still allow the attacker to compromise the database, however, this is up to the webserver security policy.

5.2 Availability

In the current implementation, availability of the server depends on the stability of the individual components (specifically, apache, mod_python and python). There webserver will restart itself on crash.

Database availability is also entrusted to the database implementation, with watchdog for restarting. Due to 3-phase commits, in case of database failure, at worst a small number of transactions may be lost.

Database integrity is ensured by updating using strict transactional approach. The database can not end up in non-consistent state (this includes failure of access logs to update), except on storage hardware failure. Regular backup schedule is handled by the database server.

6. CONCLUSION

In this paper we have presented a web-based system for medical teleconsultation and Internet-based health record that was developed for medical teleconsultation support on Croatian islands. The VP system is currently under development. The system is not limited to island-to-mainland teleconsultations but can also be used in other rural areas where specialist consultation is required. The VP system reduces risk to patients who cannot be transported due to medical problems. For patients who can be transported it reduces discomfort due to travel. VP also reduces health care costs since it eliminates the need for travel.

REFERENCES

1. Stalidis, G., Prentza, A., Vlachos, I. N., Anogianakis, G., Maglavera, S., Koutsouris, D., 2000, Intranet Health Clinic: Web-based medical support services employing XML, Proceedings of the Medical Informatics Europe 2000, IOS Press, pp. 1112-1116
2. Sussmann, H., Griebel, H., Allescher, H.-D., Egger, K., Sandschin, W., Horsch, A., 2000, The teleconsultation service ENDOTEL: Implementation and first experiences, Proceedings of the Medical Informatics Europe 2000, IOS Press, pp. 1117-1121
3. van Ginneken, A. M., Moorman, P. W., 1997 , Handbook of Medical Informatics (J.H. van Bemmel, M.A. Musen, eds.), Springer-Verlag
4. Date, C. J., 1986, An introduction to database systems, Addison-Wesley
5. Apache webserver: http://www.apache.org
6. XSL Transformations (XSLT), Version 1.0, W3C Recommendation 16 November 1999, http://www.w3.org/TR/xslt
7. PostgreSQL database: http://www.postgresql.org
8. Sablotron XSLT processor: http://www.gingerall.com
9. OpenSSL library: http://www.openssl.org
10. Trusted Computer System Evaluation Criteria (Rainbow Series: Orange Book), US Department of Defense, DOD-5200.28-STD

MEDICAL DATA TRANSFER

The Next Generation Internet (NGI) for Telemedicine Applications
Invited Paper

H.K. HUANG
Department of Radiology, Childrens Hospital Los Angeles/University of Southern California and Medical Informatics; The Hong Kong Polytechnic University, Mailstop #81: 4650 Sunset Boulevard, Los Angeles, California 90027, Hkhuang@aol.com

Key words: Internet, Internet2, Next Generation Internet (NGI), medical imaging, telemedicine, teleradiology, data security

Abstract: The next generation internet (NGI) or Internet2 is an initiative for the integration of higher speed backbone communication networks as a means to replace the current inadequate Internet for many applications including medical imaging. This presentation first reviews the current status of the Internet2 and reasons why telemedicine and teleradiology require its service. Network security and data integrity are discussed in both the global and the local network level. Some preliminary performance measurements of medical imaging applications using Internet2 in the regional and the national level, as well as comparing with other WAN technologies are given.

1. INTRODUCTION

During the past several years, Internet has encountered unexpected demands by various applications including medical imaging. The characteristics of medical imaging, especially in radiology, are different from other types of data in two aspects: its file size is very large (e.g., a mammography examination with four images is about 160 Mbytes) [1], and some image examinations require near real-time transmission for practical use [2,3]. Standard Internet with its low bandwidth, therefore has not been able to fulfill the transmission speed requirement in daily clinical applications. Due to health care reform in the U.S. during the past several years, telemedicine and teleradiology gradually become an integral part of its health care delivery system. In this respect, the problem of high-speed transmission of large volume medical images is viewed as a very critical problem [4]. The NGI initiative [5] utilizing various high-speed network backbones in the U.S including the CalREN-2 (California Research and Education Network), the vBNS (very high performance Backbone Network Service), and the Abilene to form the Internet2 infrastructure provides an opportunity for tackling this problem [6-8].

2. BACKGROUND

2.1 NGI Initiative

The NGI initiative was announced by the US federal government On October 10, 1996 based on multi-agency federal research and development programs with several major goals, among them are:

- Develop advanced networking technologies,
- Develop revolutionary applications that require advanced networking, and
- Demonstrate these capabilities on testbeds that are 100 to 1000 times faster than today's Internet.

In the health care area, telemedicine and emergency medical response team support, are some of the potential application areas for the NGI [5,11].

2.1 Internet-2

Internet2 is an academic collaborative effort to develop advanced high speed Internet technology, and to develop applications vital to research and education missions of higher education, such as telemedicine, digital libraries, and virtual laboratories that are not possible with the technology underlying today's Internet. Today over 150 U.S. universities, working together with partners in industry and government, are leading the Internet2 project. Internet2 consists of three backbones: vBNS, CalREN-2, and Abilene.

2.2 Telemedicine and Teleradiology

Telemedicine can be simply defined as the delivery of health care using telecommunications and computer technologies. Teleradiology is a subset of telemedicine dealing mainly with images and related data. There are two models in telemedicine and teleradiology; the referring physician or health care provider can either consult with individual specialists at various places through a network, or request opinions from a consolidated expert center where different types of consultation services are provided [3].

In the expert teleradiology center consultation process, three models are possible: telediagnosis, teleconsultation, and telemanagement [3,12-15]. For telediagnosis, the patient's examination results and imaging studies are done at the referring physician's remote site and data and images are transmitted to the expert center for diagnosis. The urgency of this service is nominal and turn around can take between four hours to a day. For teleconsultation, the patient may be still waiting at the examination site, while the referring doctor requests a second opinion or diagnosis from the expert center within half an hour. For telemanagement, the patient may still be on the gantry or in the examination room at the remote site, and the expert is to provide immediate management care to the patient in situ. Despite these three different models in teleradiology, the requirement of low cost, high speed NGI is obvious because of the large image file size in medical images.

3. CONNECTION TO THE INTERNET2

Currently, the Internet2 backbone is composed of several backbone networks described above which can use both ATM (asynchronous transfer mode) and PoS (packet over Sonet) technologies dependent on the locality of the backbone network. For this reason, when transmitting data through the Intenet2 backbone, the data may go through several cell-to-packet and packet-to-cell conversions. From the end user’s point of view, once the data passes through the local network gateway, there is no more local control over how the data will be transmitted through the backbone. For this reason, the total network performance will depend on the routing of the data through the Internet2 backbone as well as the routing of the data within the sending and the receiving sites. Also, data security will have two levels, one at the transmitting and receiving sites, and the other due to the Internet2 built-in security system.

Figure 1 shows an example of a connection between a local area network to the Internet2.

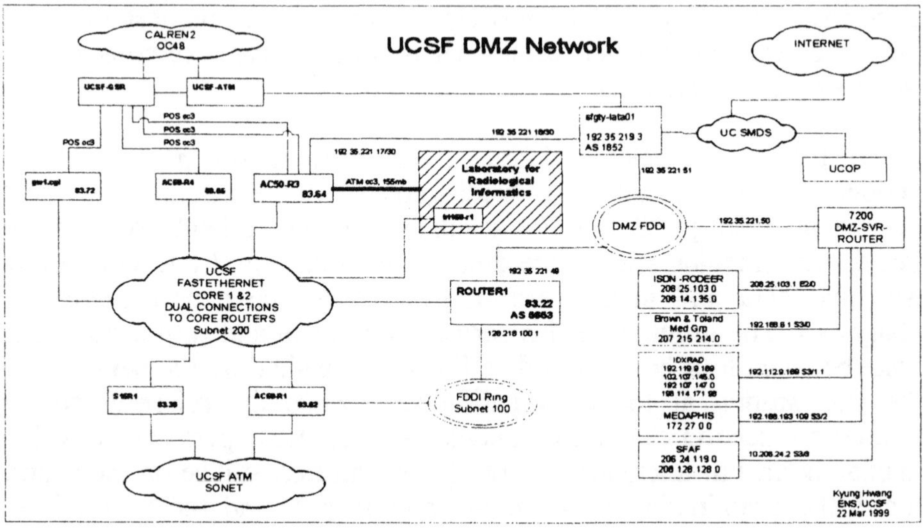

Figure 1. An example showing the connection of a LAN to the Internet2 through the CalREN-2 backbone (top left). The telemedical imaging application is at the Laboratory of Radiological Informatics (Center shaded rectangle).

3.1 SECURITY ISSUES

In order to observe data security in using the Internet2, two levels of security measures will be needed. The first level is global including the Internet2 security system and the campus/hospital network security system. Both security measures are the responsibility of the Internet administration and the campus/hospital network authority, respectively. In this presentation, we focus our attention at the local level, that is, the responsibility of the end user.

In teleradiology, the question raised often is how authentic is the medical image after it is transmitted through an open network. Generally, trust in digital data is characterized in terms of privacy, authenticity, and integrity of the data. Privacy refers to denial of access to information by unauthorized individuals. Authenticity refers to validating the source of a message. Integrity refers to the assurance that the data was not modified accidentally or deliberately in transit. Thus, privacy is the responsibility of the Internet2, the campus network authority, and the end user's private network manager, and in that order; authenticity and integrity are the responsibility of the end user.

In a teleradiology operation, the end user is responsible for maintaining its private network and its network and data security. First consider the local network security, if the network used is ATM, an ATM hardware firewall will be needed [9]. A software firewall will be insufficient due to the fact that software encryption and decryption will delay the transmission speed for telemanagement application. The encryption delay in current ATM OC-3 cell-based hardware firewall is on the order of 32 microseconds, well acceptable for telemanagement applications [16]. Once the data is encrypted via the ATM firewall, data is secured up to the point specified by the hardware. If the local network is PoS, network firewall is still using software, almost real-time hardware firewall is not yet available.

The end user is also responsible for guaranteeing the integrity of the image data. This level of security is mostly done based on the concept of digital signature of the image using some existing cryptography algorithms while the image is being generated. The digital signature based on certain characteristics of the medical image and related patient data will provide additional protection and assurance of data authenticity and integrity that are not available in the ATM or PoS firewall. In general the digital signature concept consists of four steps:

Image pre-processing: To segment objects of interest from background and extract patient information from the DICOM (Digital Imaging and Communication in Medicine) image header.

Image digest: To compute an image digest (or digital signature) of the image based on its characteristics using the MD5 digest or other algorithm [10].

Data encryption: To produce a digital envelope containing the encrypted image digest and corresponding patient information from the image header.

Data embedding: To embed the digital envelope into the image. This is done by replacing the least significant bit of a random pixel of the image by one bit of the digital envelope bit stream and repeating for all bits in the bit stream.

4. DISCUSSION

Internet2 consists of several existing high-speed network backbones including vBNS, CalREN-2, and Abilene. These backbones use cell-based ATM and packet-based PoS technologies. In order for the end user to connect to and take advantage of the Internet2, a major overhaul of the local network architecture is necessary if the specifications of the local area network are not sufficient to match the Internet2 perfromance. Compatible performance network routers and packet-cell converters should be used in the new architecture.

Internet2 will provide necessary data security at the global level. But the data security at the local level is the responsibility of the end user. In this regard, the end user has to provide its own private network security as well as data integrity. If the local network used is ATM, the network security can be taken care of by hardware ATM OC-3 firewall which has an encryption delay time of 32 microseconds, acceptable for most medical imaging applications. The data integrity relies on software encryption during the formation of the image, still a very time consuming process. For the PoS packet protocol firewall, since hardware is not yet available, current technology relies on software. In order to speed up the encryption/decryption process, special processors in the main switches are needed. Research and development in hardware packet protocol firewalls is currently an urgent goal for the network community.

In medical imaging, average file size per examination is in the neighborhood of 10-100 Mbytes, and getting bigger. For this reason, current wide area network technology is either too slow or too costly for telemedicine and teleradiology applications involving images. Internet2, fast and inexpensive, is the potential solution for the rising demands of these applications.

Performance of using Internet2 for medical imaging applications depends on two factors. First, the Internet2 and the local connection including data paths, routers, and communication protocols in both the global and the local network architecture will affect its performance. Second, the performance is application dependent, especially in the health care field. The recently national NGI projects on health care applications should shed some light in this respect (see Acknowledgement). We expect the performance of Intenet2 to improve substantially during the next several years compared with today's measurements.

ACKNOWLEDGEMENT

This work is partially supported by contracts N01-LM-6-3547 and N01-LM-9-3538 from the National Library of Medicine, and DAMD17-99-P-3732 from the U.S. Army Medical Research and Material Command.

REFERENCES

1. Huang HK, Lou SL. 1999. Telemammography: A Technical Overview. *RSNA Categorical Course 1999,* 273-281.
2. Huang HK, 1997. Telemedicine and Teleradiology Technologies and Applications. *Min Invas Therapy*. Vol 5/6, 387-392.
3. Huang HK. 1996. Teleradiology Technologies and Some Service Models. *Comp. Med. Imaging & Graphics*. 20:59-68.
4. Huang HK, Arenson RL, Dillon WP, Lou ASL, Bazzill T, Wong AWK, 1995. Asynchronous Transfer Model(ATM) Technology for Radiologic Communication. *Am. J. Roentgenology* 164:1533-1536.

   Http://www.ngi.gov
5. Http://www.internet2.edu
6. Http://www.ngi.gov
7. Http://www.calren2.net
8. Celotek Corporation, " Encrypting ATM Firewall", http://www.celotek.com
9. XQ Zhou, SL Lou, and HK Huang, 1999, Authenticity and Integrity of Digital Mammographic Images, *Proc SPIE, Med. Imaging*, 3662-16.
10. NGI Implementation Plan, http: www.ccic.gov/ngi/implementation.
11. Zhang J, Stahl JN, Huang HK, Zhou X, Lou, SL, Song KS, 1999. Real-time teleconsultation with high resolution and large volume

medical images for collaborative health care. *IEEE Trans Inform Tech Biom.* (accepted).

12. Huang HK. 1999. PACS: Basic Principles and Applications. John Wiley's & Sons, NY, NY. PP.521.

13. Lou SL, Sickles EA, Huang HK, Hoogstrate D., Cao F., Wang J., and Jahangiri M., Full-field direct digital telemammography: technical components, study protocols and preliminary results. *IEEE Trans. Inform. Tech. in Biomedicine,* Vol. 1, No. 4, 270-278, 1997.

14. Huang HK, Wong AWK, and Zhu X. Performance of Asynchronous Transfer Mode (ATM) Local Area and Wide Area Networks for Medical Image Transmission in Clinical Environment. *J Comp Med Imag & Graphics* 21(3), 165-173, 1997.

15. Celotek Corporation, " Cellcase product line", http://www.celotek.com.

VESIMA-MED

Energy and Data Transmission

[1]F. PRÄMAßING, M.SCHMIDT, [2]V. BIEFELD, J. BINDER
[3]B. CLASBRUMMEL AND B. JETTKANT
[1]EPOS GmbH & Co. KG., an Infineon Technologies Company, Duisburg, Germany: [2]Institute for Microsensors, -Actuators and –Systems (IMSAS), University of Bremen, Germany: [3]Berufsgenossenschaftliche Kliniken "Bergmannsheil", Bochum, Germany

Key words: VESIMA-MED, implantable monitoring system, capacitive acceleration sensor, readout circuit, telemetry

Abstract: This paper presents a micro-movement long-term monitoring system optimising the healing process of fractures. It focus in particular on the energy and data transmission path of the system. Continuous and accurate monitoring near the fracture zone of bones requires an implantable and low power consumption device. Additionally, the implant needs to be small in size. To serve as a long-term implant, no batteries are used due to their limitations in lifetime and power. Thus, the system is inductively powered based on the inductive coupling between an external and an internal coil. Monitoring is done by a miniature, capacitive accelerometer. Using a telemetric link the measured data stream is transmitted from the implant to a microcontroller, where the data is analysed.

1. INTRODUCTION

In orthopedic or traumatological research, and clinical applications, the beneficial effect of applied micro-movements on fracture healing has been demonstrated [1,2]. With indirect micro-movement monitoring systems (e.g. via bending of the fixator) the fracture motion characteristics can only be determined approximately. Additionally, pin loosening may cause errors in data acquisition. The

"VErtical Structured Intelligent Micro Accelerator with application in MEDicine" (VESIMA-MED) is meant to be a miniaturised measurement device for clinical applications using a non invasive concept for monitoring amplitude and acceleration. It applies precise micro-movements in order to optimise the healing process of the bone. Thereby, it is expected to decrease the duration of fracture healing as well as rehabilitation, and possibly reduce the rate of complications encountering.

1.1 Biocompatible Packaging

The application of implantable medical sensor systems imposes some important restrictions to the system specifications in the fields of quality and reliability. The VESIMA-MED system comprises several single modules like acceleration sensor, energy supply, telemetry etc. which are encapsulated. As the system is meant to be implanted biocompatibility and full hermetic sealing over decades of years are essential requirements of the packaging. Since the coating material moves and swings together with the implanted system the encapsulation does not influence the functioning of the system. With regard to interactions between the sensor system and other diagnostic or therapeutical devices a careful design is necessary.

2. SYSTEM DESCRIPTION

2.1 Data and Energy Transmission

The integration of sensors and electronics offers several significant advantages especially in medical applications, such as small-size and low-weight, close physical proximity of sensors and readout. Due to the low power consumption CMOS electronics is ideal for the use in implanted systems. The implant receives energy through wireless RF telemetry and incorporates sensor, readout circuit, and telemetry chip. The external unit comprises the external transmitter and microcontroller (Fig. 1). The system works as follows: Capacitive changes of the sensor are converted into a digital data stream by the readout circuit. This data stream is transmitted by the telemetry chip to the external transmitter. At the receiver side a data recovery circuit within the external transmitter transforms the transmitted data stream into signals suitable for further digital processing. The microcontroller is used either for monitoring the data set or controlling an actuator.

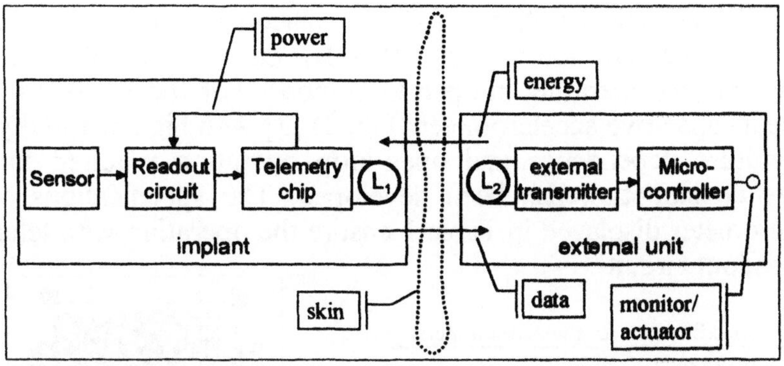

Figure 1. Data and energy transmission using wireless RF telemetry.

The readout circuit (Fig. 2) comprises a charge to voltage (C-V) converter, a differential amplifier performing an offset calibration and an amplification and a sigma-delta converter. The output signal is a one bit stream, representing the analog input signal. In order to make the system less sensitive to distortion both the sigma-delta modulator and the amplifier using a fully differential technique. Furthermore, the one-bit data stream is less sensitive to noise coupling and provides sensor data with a higher resolution than analog circuits.

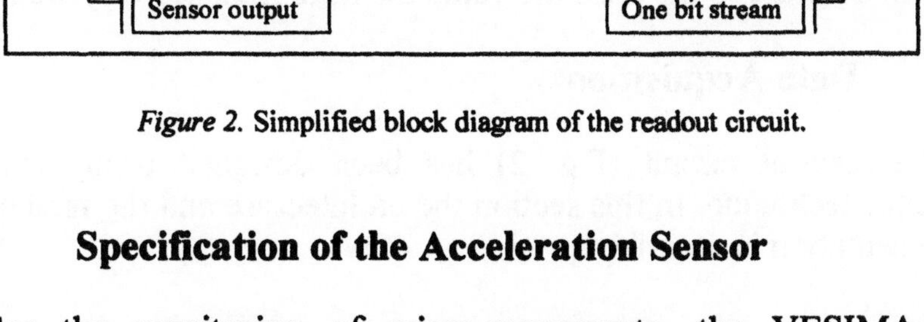

Figure 2. Simplified block diagram of the readout circuit.

2.2 Specification of the Acceleration Sensor

For the monitoring of micro-movements, the VESIMA-MED device has to detect accelerations up to 3 g with a resolution of 10 mg.

Available acceleration sensors fabricated in surface micro machining can meet this specification by using closed-loop operation. Due to the limited energy budget for the implant these close-loop devices are not suitable for telemetric applications. Microsystems technologies such as chemical mechanical polishing (CMP), silicon direct bonding (SDB) and deep reactive ion etching (DRIE) enable the fabrication of accelerometers operating in open-loop mode. For the VESIMA-MED device a capacitive accelerometer (Fig. 3) [3] with high sensitivity, low temperature dependence and open-loop operation required for low energy consumption has been developed. The specifications of the accelerometer displayed in Table 1 ensure the operation with telemetry and readout circuit.

Table 1. Specifications for the accelerometers

Measurement range	± 3 g
Resolution	10 mg
Total capacitance	< 8pF
Capacitance range	± 300 fF FS
Bandwidth	≤ 200 Hz
Linearity	< 1 %
Natural frequency	> 1 kHz

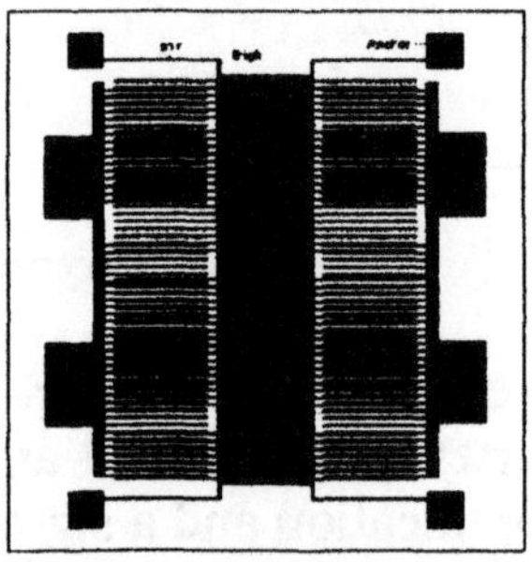

Figure 3. Layout of the designed accelerometer.

3. SYSTEM DESIGN

The implementation of resistors and capacitors using integrated circuit technology is often very difficult due to their required area. Additionally, tolerances of ± 30% occur. Using a switched capacitor (SC)-realisation, resistors can be substituted very easily. SC-design makes it possible to reduce the resistor parameter to the capacitance ratio. In contrast to the absolute value the ratio matches very well.

3.1 Data Acquisition

The readout circuit (Fig. 2) has been designed using switched capacitor technique. In this section the architecture and the final circuit implementation is described.

3.1.1 SC Capacitor-Voltage Converter

Figure 4 shows a simplified schematic of the sensor and the switched capacitor C-V converter.

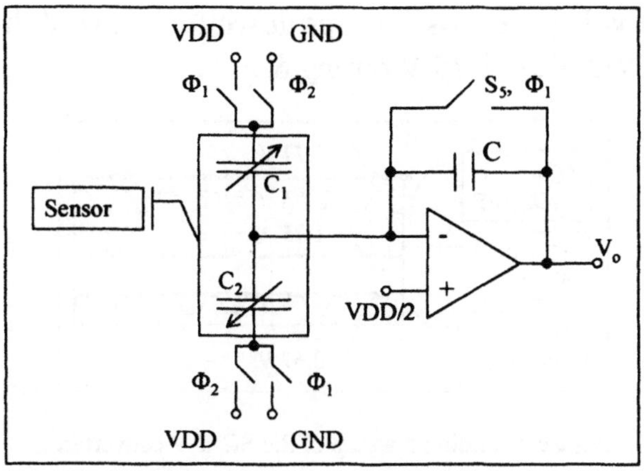

Figure 4. Simplified schematic of the sensor and the switched capacitor C-V converter.

The sensor symbolised by C_1 and C_2 acts as a differential capacitor. It provides the CV-converter with the differential capacitance $\Delta C = C_1 - C_2$ resulting from the occurrence of an acceleration near the fracture zone. The amplifier has been designed and implemented according to E. Vittoz [4] and delivers a voltage proportional to ΔC. The C-V converter works as follows: During ϕ_1 the capacitance C placed within the feedback loop is discharged by switch S_5. The charge corresponding to the differential capacitance
$\Delta C = C_1 - C_2$ transferred from the sensor onto C during ϕ_2 is given by

$$Q = (VDD\text{-}GND)\cdot(C_1 - C_2).$$

Assuming an infinite amplifier gain the output of the
C-V converter is equal to

$$V_o = (VDD\text{-}GND)\cdot(C_1 - C_2)/C + VDD/2.$$

In order to analyse the performance of the C-V converter the voltage timing diagram (Fig. 5) as a result of the charge flow from the sensor to the capacitance C is displayed. Three different values of ΔC representing three different accelerations are considered. According to Table 1 the differential capacitance ΔC = +368.76 fF corresponds to an acceleration of 3.7 g approximately. During clock cycle ϕ_2 the output V_0 goes from VDD/2 = 1.75 V up to 2.77 V and goes back to VDD/2 during ϕ_1. For ΔC = 0 fF the output of the amplifier remains constant at 1.75 V. For an acceleration corresponding to ΔC = -368.76 fF the voltage across C drops down to 0.67 V during ϕ_2 and goes back to VDD/2 = 1.75 V during ϕ_1.

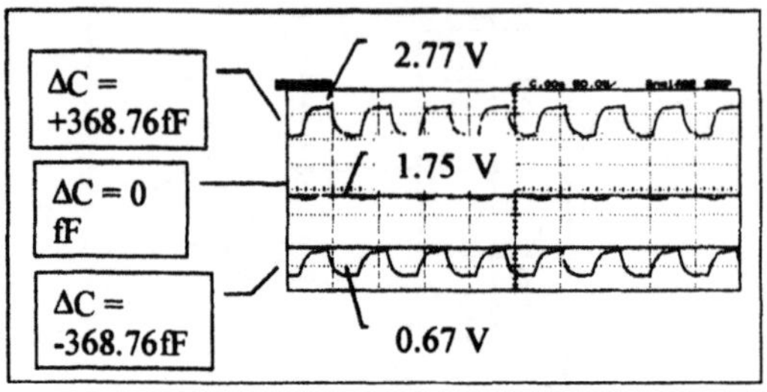

Figure 5. Timing diagram of the SC C-V converter.

3.1.2 Fully Differential Switched-Capacitor Amplifier

In order to minimise lithography errors the capacitance C within the feedback loop of the CV-converter should be at least 1 pF. This leads to a limitation in the gain factor and can be compensated by adding a second amplifier to the signal path. The design of this amplifier is based on a fully differential technique [5] which makes the signal path less sensitive to low frequency or power supply noise. A realisation of the SC-amplifier is displayed in Figure 6. A time–domain measurement demonstrating the principle of operation is given in Figure 7. In contrast to the CV-converter the SC-amplifier has two output signals balanced with respect to VDD/2. At the output nodes appear the voltages equal to

$$V_{o+} = ((V_o - VDD/2) / 2) (C_a / C_b) + VDD/2$$

$$V_{o-} = -((V_o - VDD/2) / 2) (C_{a'} / C_{b'}) + VDD/2$$

Where $(C_a / C_b) = (C_{a'} / C_{b'})$ are tunable in the range of 1 to 4.5.

The output voltages V_{o+} and V_{ot-} do not change significantly during the clock phase ϕ_2 since a sample and hold circuit is placed within the feedback loop of the amplifier. Assuming a capacitance change of ΔC

= +368.76 fF V_{out-} goes up to +2.56 V while V_{o+} decreases. This results in a differential output of

$V_{o-} - V_{o+}$ =2.56 V –0.95 V = 1.61 V according to the gain of

$(C_a / C_b) = (C_{a'} / C_{b'}) = 1.5$.

Again as long as ΔC = 0 fF the output voltages remain unchanged. ΔC = -368.76 fF leads to the output voltages V_{o+} = 2.6 V and V_{o-} = 0.88 V respectively.

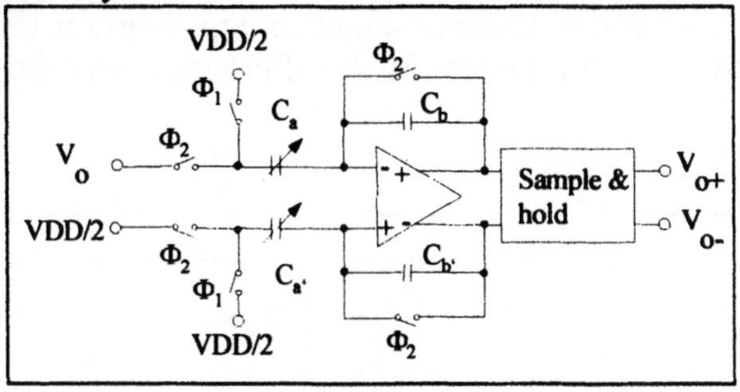

Figure 6. Simplified schematic of the SC differential amplifier.

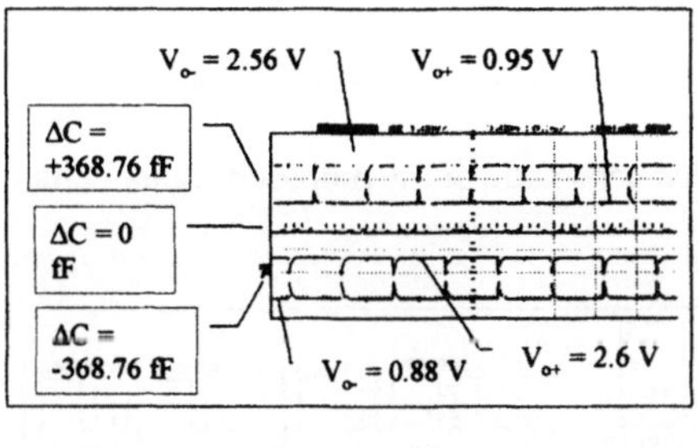

Figure 7. Timing diagram of the SC differential amplifier

3.1.3 Second-Order ΣΔ Modulator

Converting the analog signal into a digital signal the sensitivity to distortions of the sensor system is reduced. A sigma-delta modulation is a well-suited way to perform high-resolution A/ D conversion of relatively low bandwidth signals. Since the analog input signal is sampled at a frequency which is much higher than the Nyquist rate high frequencies and quantization noise above the desired signal bandwidth can be removed by a digital low-pass filter. The required resolution realised by a first order sigma-delta modulator needs oversampling rates which are beyond the limits of the IC process. Reducing the oversampling ratio of a sigma-delta converter causes a corresponding reduction in the dynamic range. Fortunately, this reduction can be compensated by increasing the order of the modulator. A second-order ΣΔ modulator contains two integrators in the forward path and offer higher resolution than a first-order ΣΔ modulator at the same oversampling rate. Modulators with higher resolution suffer from instability and cause an increase in power consumption. Figure 8 shows a schematic of the designed ΣΔ modulator [6], with the corresponding timing diagram (Fig. 9). The designed converter comprises fully differential switched-capacitor integrators.

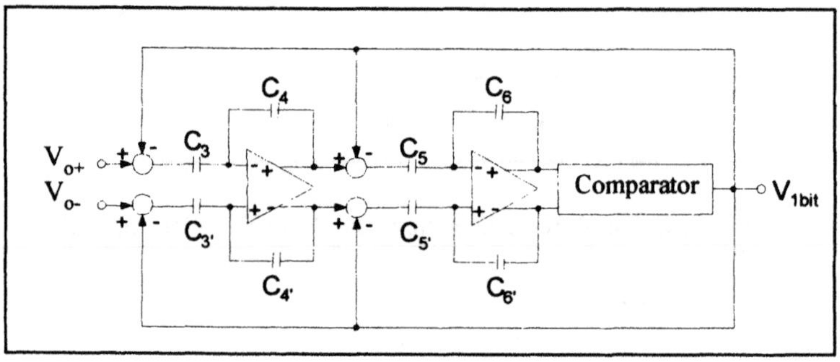

Figure 8. Simplified schematic of the proposed sigma-delta modulator.

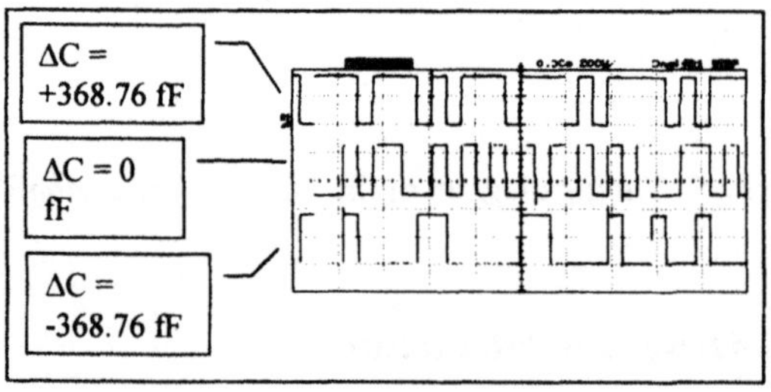

Figure 9. Timing diagram of the sigma-delta modulator.

3.2 Inductive Power Link

Figure 10 gives an overview of the energy transmission path. The telemetry chip can be divided into the clock, energy, and the data generation path. The RF field is generated by the energy transmitter coil L_2. The coil L_1 receives the RF energy of the external energy transmitter for powering the readout circuit. The system clock for the readout circuit is extracted from the external RF field by a clock driver. A CMOS rectifier consisting of Schottky-diodes generates a suitable bias for the voltage generator providing both the voltage supply for the analog and the digital components. The modulated sensor information is transmitted to the external transmitter by $L_{1'}$. If the transmission gate T is closed additional energy from the RF field is absorbed resulting in a change in the voltage amplitude across the inductance L_2. Using this voltage drop the transmitted data is detectable.

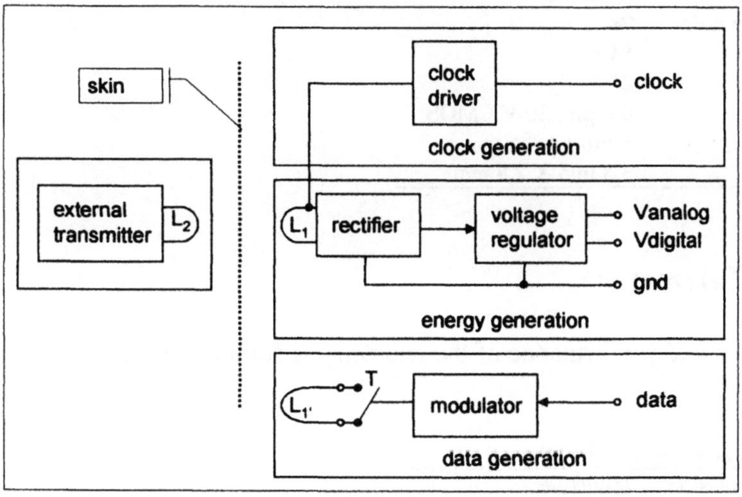

Figure 10. Energy transmission path including external transmitter and telemetry chip.

4. CHIP PERFORMANCE

4.1 Readout Circuit

The characteristics of the readout circuit are displayed in Table 2.

Table 2. Characteristics of the readout circuit

Voltage supply	$V_{DD} = 3.5$ V
Quiescent current	0.5 mA ($V_{DC} = 3.5$ V)
System clock	15.625 kHz
Signal bandwidth	≤ 200 Hz
Oversampling ratio	78
Resolution	54 dB
ΣΔ modulator	2nd
Technology	0.8 µm, 20 V CMOS
Modulation	1-bit-data stream
Chip area	3.8 mm X 2.8 mm

4.2 Telemetry

The measured performance of the telemetry is summarised in Table 3.

Table 3. Characteristics of the telemetry

Energy transmission principle	Inductive
Voltage supply	$V_{DD} = 3.5$ V
RF frequency	4 MHz
Quiescent current	130 µA
Data rate	125 kbaut
Technology	0.8 µm, 20 V CMOS
Chip area	2.7 mm X 2.2 mm

5. CONCLUSION

An implantable micro-movement monitoring system for optimising the healing process of fractures has been presented. The monitoring of the micro-movements is realised by applying a miniature accelerometer, placed direct at the bone or at the bone plates. Data from the implant is wirelessly transmitted and analysed by a microcontroller. The system receives energy using an RF telemetry.

ACKNOWLEDGMENTS

This work was financially supported by the BMBF. The authors wish to thank all of the partners of the VESIMA-MED consortium for their contribution to this work.

REFERENCES

1. Goodship, A. E., and Kenwright, J., 1985, The influence of induced micromovement upon the healing of experimental tibial fractures. J Bone Joint Surg [Br], 67(4), 650-655
2. Clasbrummel, B., Muhr, G., and Goodship, A. E., 1997, Zeitabhängige mechanisch Stimulation in der enchondralen Frakturheilung beim Schaf. Biomed. Tech. Band 42, 521-522
3. O. Lüdtke, V. Biefeld, A.Buhrdorf, J. Binder, October 1999, Laterally driven accelerometer fabricated in single crystalline silicon, in Sensors and Actuators A 82, p.149 –154,
4. M. Degrauwe and e.Vittoz, June 1985, A Micropower CMOS-Instrumentation Amplifier in IEEE J. Solid State Circuits, vol. sc-20, no. 3,
5. M. Banu and J.M. Khoury, December 1988, Fully differential operational amplifiers with accurate output balancing, IEEE J. Solid State Circuits, vol. 23, no. 6,
6. Steven R. Norsworthy, R. Schreier, Gabor C. Temes, USA1997, Delta-Sigma Data Converters Theory, Design, and Simulation, IEEE Press.

STANDARDISATION AND DATA COMPRESSION

Applicability of Lossy Image Data Compression in Tele-Medicine

Mathematical basis, technical realisation and medical evaluation

[1]WERNER STEFANON, [2]DOROTHEE P. AUER, , [1]LUDWIG M. AUER
[1]Department of Medical Telematics and Robotics, Institute of Applied Sciences in Medicine (ISM), AUSTRIA, Jakob-Haringer-Strasse 3, A-5020 Salzburg
[2]Arbeitsgruppe NMR, Institut für klinische Psychiatrie, Max-Planck-Institut München, Kraepelinstraße 10, D-80804 München

Key words: Medical image data compression, teleradiology, Compression automation

Abstract: Visual Aim of this study was therefore to investigate whether mathematical methods could help to develop a tool for automated image compression based on clinically acceptable rates for JPEG and Wavelet compression for different imaging modalities. A reproduceable parameter for description of image quality, the signal to noise ratio was calculated on each image (Peak signal to noise ratio, PSNR). 852 JPEG and 640 Wavelet-compressed MR images in series 1 and 24 MR-images, 8 MR-angiograms, 21 renal MR-angiograms and 6 conventional myelograms, each at 10 compression rates with JPEG and Wavelet in series 2 were used for 2 strategies for semi-quantitive evaluation. The results indicates that the signal to noise ratio is not an adequate approach for subjective evaluated image quality and can therefore not be used for an automated compression process within a transmission software. Medical evaluation as a gold standard indicated, that compression up to 20:1 are acceptable. Moreover, Wavelet compression seemed to be superior for renal MR-angiograms and x-ray-myelograms. New generation of teleradiological software will include a tool for manually and visually controlled image data compression.

1. INTRODUCTION

The range of acceptable compression rates with different methods and on different types of medical images differs between imaging modalities.

Recent studies have been performed in cardiology (9 – 13), neonatal intensive care (7), mammography (5, 14), dermatological imaging (15) and chest X-ray (16). Acceptable compression rates in neuroimaging are widely unknown.

The opinion expressed by some authors, that modern compression algorithms do not effectively reduce image quality, even when subjected to compression ratios up to 30:1 (4), does not necessarily apply to all different imaging modalities nor to different compartments within single images.

There are mainly two different ways of compression of digital image data. One can reconstruct the original image numerically identical (loss-less compression) and the other one tries to take advantage from psycho-visual effects to optimise compression results but looses certain image information (lossy compression).

Lossy methods prepare image data in a way that loss-less compression algorithms (based on the removal of redundancy) can deploy their effect more efficient.

In a nutshell, JPEG quantizes high frequency parts of the image expecting to remove them totally and provide a lot of identical symbols (namely Zeros) to the redundancy removing block (see figure 1). Decompression runs the other way round. It is a symmetric algorithm, the same calculation steps/time is needed to compress and decompress an image. The source image is blocked to 8x8 image parts to speed up compression.

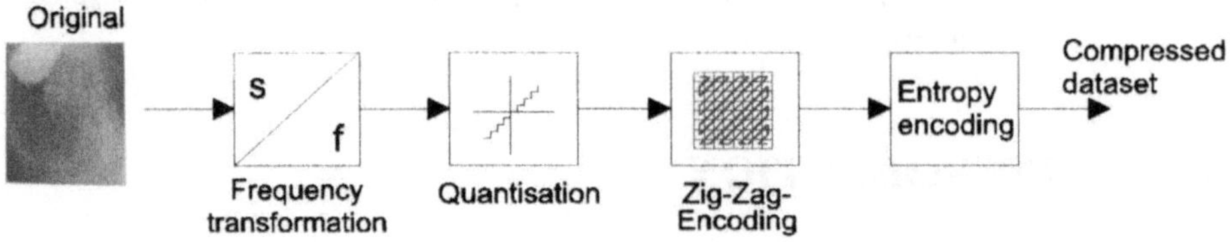

Figure 1: JPEG-encoder

The Wavelet compression method works globally on the image and tries to replace sequences of pixels with recursively assembled small parts of waveforms (wavelets). There are a quite substantial number of wavelets (3) in existence but the currently used is the Daubechi's wavelet and its derivatives (see figure 2).

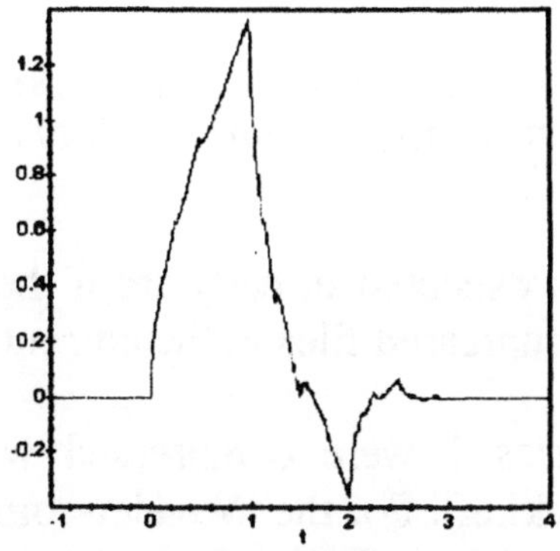

Figure 2: Daubechi's wavelet

In the present study, series of different medical images were compressed and presented to clinicians for evaluation in order to identify, if any, acceptable methods of image compression as well as allowable compression rates.

Two lossy algorithms were analysed, JPEG (2) and Wavelets (implemented in JPEG2000).

2. METHODS

2.1 Compression process

Series of medical images were compressed with both algorithms at different compression ratios. The different image processing programs do not result in a substantial difference in image quality compared to compressed file size (objective measurement with PSNR, see 2.3). Therefore, Matlab® (4) was used to partly automate the compression process.

```
%JPEGger
%By Werner Stefanon
function jpegger (path);
[a,map_a]=imread(path,'bmp');
for i=10:10:100
   i
   fname=strcat(path,sprintf('-jpeg%3d.jpg',i));
   imwrite(a,fname,'jpg','quality',i);
end
```

The script above is executed directly from the Matlab® shell and produces ten JPEG-compressed files in the current directory.

The images in series 1 were compressed with a Haar-Wavelet algorithm (simple blocking). For the Wavelet-compression in series 2, MT-Wice® Photo from Mevis Technology was used. Ten compressed images with linear progressing quality settings were generated and stored.

2.2 Imaging modalities and evaluation

All images were randomly listed and presented to clinicians in order to determine acceptability of image quality. If the distortions were beyond acceptability, the image was tagged as unacceptable, otherwise as acceptable or as poor (however not distinguishable, whether primarily poor image quality or result of compression). The results were statistically evaluated to see the maximum compression ratio accepted with different image modalities (different image capturing modalities).

Loss-lessly compressed images and original images were also mixed into the series.

Analyses at different levels of compression were performed on two series of images:

Series 1

- 71 brain MR images were used for compression at 12 different levels, resulting in a set of 852 (JPEG) images;
- 64 brain MR images were used for compression at 10 different levels, resulting in a set of 640 (Wavelet) images.

- Repeated analysis was performed by two investigators. Image quality was ranked as acceptable or unacceptable (see above).
- Inter- and intra-observer variability were calculated and statistical evaluation was performed using Kappa-statistics. All test runs were compared using the individual results of each evaluator and averaging the results of all evaluators.

Series 2

- Various numbers of brain MRI, angiograms, chest X-rays and bone X-rays were compressed at 10 different levels with both methods.
- The resulting file of 967 images was analysed as a preliminary series of an ongoing study.

2.3 Mathematical evaluation

To number the statistical difference between one original and its compressed image, the PSNR (Peak Signal to Noise Ratio, see Equation 1) was calculated. This number can correspond with the psycho-optical impression of human evaluators.

$$PSNR = 10 \cdot \log_{10} \frac{\left(\max\left[\max(f(x,y))\ \max\left(\tilde{f}(x,y)\right)\right]\right)^2}{\frac{1}{N \cdot M} \cdot \sum_{i=1}^{N} \sum_{j=1}^{M} \left(f(x_i, y_j) - \tilde{f}(x_i, y_j)\right)^2}$$

Equation 1: PSNR-Calculation

Where N . . . x-resolution [Pixel],
M . . . y-resolution [Pixel],
f . . . original image,
~f . . . compressed image.

The calculation of the PSNR was also done with a Matlab® script, but with an extended the functionality to a graphical user interface (see figure 3).

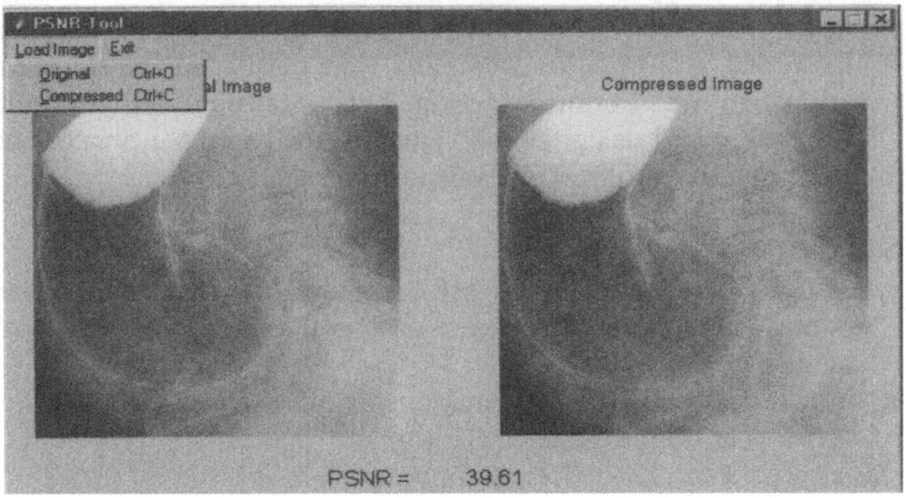

Figure 3: GUI of the image compression tool

3. RESULTS

Series 1:
On the average, Wavelet compression produced less visible artefacts than JPEG at comparable compression rates (example is shown in figure 4).

With increasing compression ratio the PSNR degrades, starting at an infimum (identical images) and going down to a few decibels. It shows that the image quality from a mathematical point of view is better when using a wavelet compression algorithm. It keeps the image quality far into high compression ratios where JPEG degrades rapidly (above 60:1).

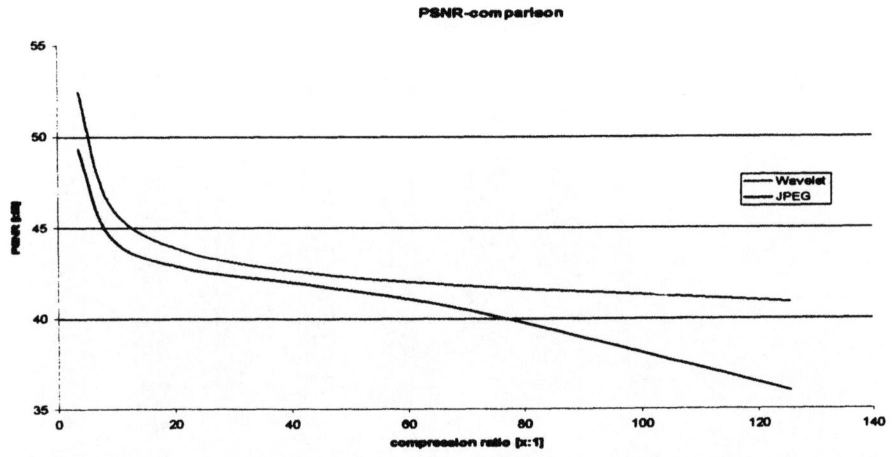

Figure 4: Comparison of JPEG and Wavelet compression The x-axis shows the compression ratio and the y-axis the PSNR. A higher PSNR value indicates better image quality.

The following graph (figure 5) shows the distribution of the three different groups of images with green (quality is acceptable), yellow (quality is borderline) and red (unacceptable) with increasing compression / decreasing quality. For example, quality setting 200 (third bar from the left in figure 5) is equivalent to a compression ratio of 6,44 to 1.

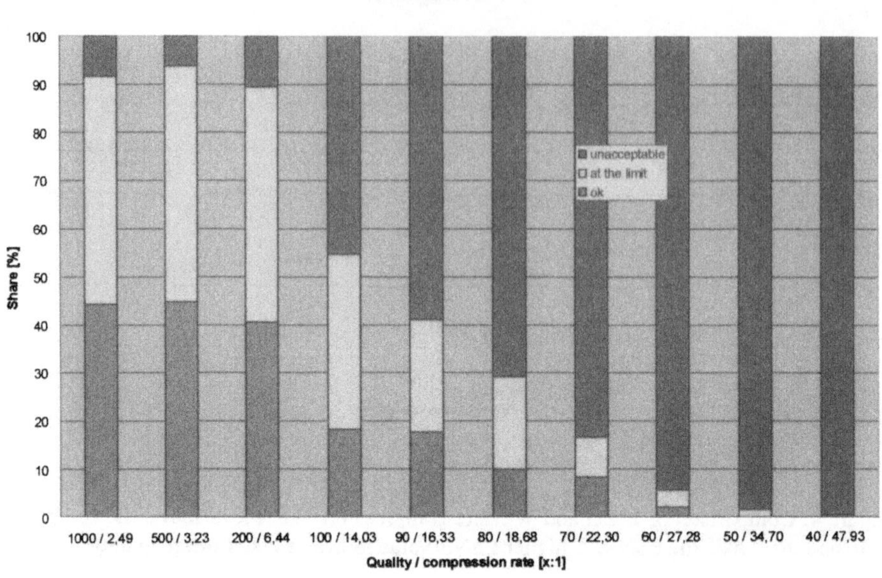

Figure 5: Evaluation of images after the 10 degrees of wavelet compression (x-axis), ordinate: PSNR (peak signal to noise ratio)

With the algorithm used in this series , compression up to 6:1 did not have a recognisable effect on the quality of MRI of the brain.

The analysis of the inter-observer variability shows a value of 0.506 with the JPEG-algorithm and a substantially higher one, 0.704, with the wavelet algorithm. Since this number is very different comparing the test runs (0.378 first run vs. 0.704 second run) a clear learning effect can be seen.

High deviations between the investigations of one evaluator and between evaluators show a difference in the subjective impression of image quality.

Series 2:
JPEG compression of high quality images up to 30:1 and Wavelet compression up to 60:1 did not influence image quality of any of the investigated imaging modalities (see figures 6, 7).

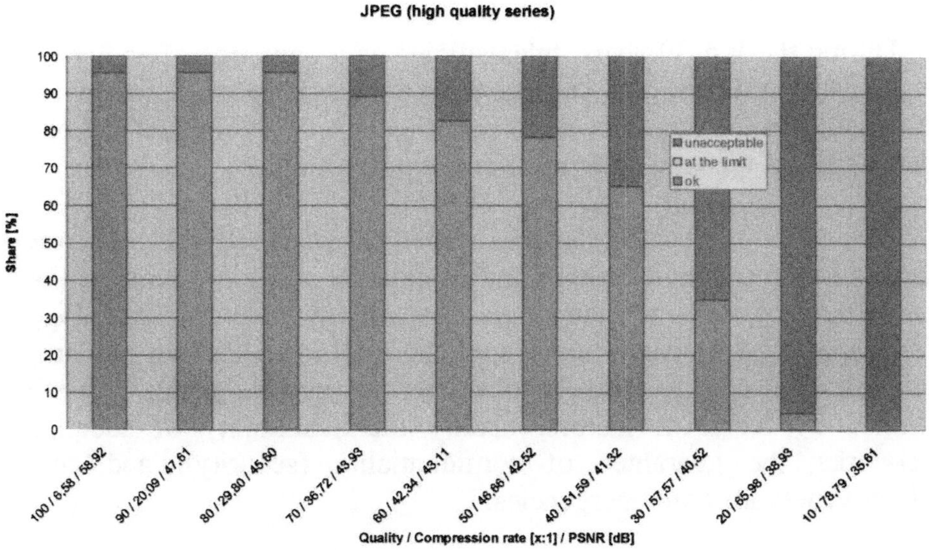

Figure 6: High quality images with JPEG-compression

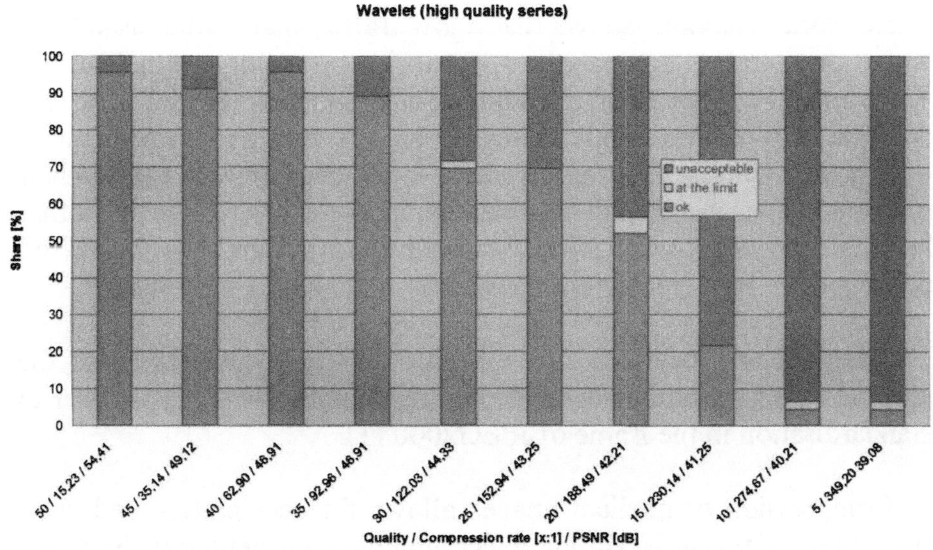

Figure 7: High quality images with Wavelet-compression

4. DISCUSSION AND CONCLUSION

During the last 30 years, telemedicine as a new area of activity in healthcare systems has undergone several changes of paradigm, depending on the rapid development of technical possibilities and also depending on the viewpoint of different user groups.

One of the core interests from a medical perspective has become the exchange of single images and of datasets via telecommunication networks (e.g. tele-radiology, tele-dermatology, tele-ophthalmology, tele-pathology). However, wide application of inter-hospital exchange as well as exchange between physicians in private practice is still limited by technical factors, mainly the availability of adequate networks, the guarantee of confidentiality (security), and cost-effectiveness for non-emergencies.

Medical image data transmission has to be accurate, safe and cost-effective . For wide acceptance of this area of tele-medicine, fast networks and low error rates are mandatory (5, 6).

Network behaviour shows that a low traffic load results usually in better error rates, compared to networks operating at their limit. Therefore, the amount of transmitted data should be reduced to the acceptable minimum. Medical image datasets such as CT, MR, or X-ray images with high resolution and several slices must not loose their relevant information from the medical point of view. On the other hand, low or no compression would lead to a traffic overload and unacceptably high transmission costs (8).

International standards are existing or in development for the analysed compression algorithms. As an example, wavelets are near to standardisation in the frame of JPEG2000 (1).

Compression of medical images allows for acceleration and cost-reduction of transmission via telecommunication networks. State-of-the art transmission only allows loss-less compression. For applications like emergency treatment, the image quality does not need to be perfect but sufficient for the purpose.

Image compression up to 40:1 (JPEG) was reported to result in sufficient image quality in dermatology (15). However, caution has been recommended in some applications. In cardio-angiography, JPEG-compression up to 6:1 was found equivalent to original images, whereas higher compression rates induced major reductions in diagnostic accuracy (9 - 11).

Concrete information on clinically allowable compression ratios for several imaging modalities is scarce or even unavailable in literature.

Results from the present study indicate that in the field of brain MRI compression ratios of 30:1 using JPEG and 60:1 using wavelets are acceptable.

These preliminary series indicate, mainly from the field of Neuroimaging, that compression of digitalised medical images such as MRI is possible to a considerable degree without recognisable loss of quality from a medical perspective. This result is biased, however, by the fact that unselected series of images of variable quality were used in series 1.

Moreover, the second series in progress on high quality images and from different imaging modalities shows, that considerable compression rates may be acceptable for a scenario of clinical decision making. However, evaluation of different degrees of allowable compression between various imaging modalities require further studies on larger series of images.

The aspect of transmission speed is evident. JPEG2000 (based on wavelet compression) might be implemented in following DICOM-standards which allow currently (version 3) JPEG compression. With this option at hand and proper guidelines for compressing a variety of image modalities, costs could be reduced to a considerable extent and transmission/consulting in telemedical scenarios may be accelerated.

For future tools in teleradiology, image compression will therefore be a valuable adjunct, provided the condition is fulfilled that the sender is informed about reasonable compression rates at large, and able to control the quality of transmitted images.

REFERENCES

1. ISO/IEC JTC1/SC29 WG1, JPEG 2000, M. Boliek, C. Christopoulos and E. Majani, 16th March 2000, *JPEG 2000 Part I Final Committee Draft Version 1.0*
2. *JPEG*, http://www.jpeg.org/public/jpeghomepage.htm
3. Amara Graps, 1995, *An Introduction to Wavelets*, IEEE Computational Science and Engineering, CA
4. *Matlab®*, http://www.mathworks.com
5. Mitra S, Yang S, Kustov V., *Wavelet-based vector quantization for high-fidelity compression and fast transmission of medical images*, J Digit Imaging. 1998 Nov;11(4 Suppl 2):24-30
6. Giacomuzzi SM, Springer P, Stoger A, Dessl A, Buchberger W, Bodner G, Bale R, Schreder JG, Gell G, Jaschke W., *The Austrian Academic Computer Network and its usefulness for teleradiology*, J Telemed Telecare. 1998;4 Suppl 1:41-2.
7. Parisi SB, Mogel GT, Dominguez R, Dao H, Cramer TJ. *The effect of 10 : 1 compression and soft copy interpretation on the chest radiographs of premature neonates with reference to their possible application in teleradiology*, Eur Radiol. 1998;8(1):141-3.
8. Bruckmann A, Uhl A, *Ein Vergleich von Wavelet und JPEG basierten selektiven Methoden im Bereich der medizinischen Bildkompression*, Forschungsinstitut für Softwaretechnologie (RIST++), Paris-Lodron-Universität Salzburg, Proceedings des Workshops Bildverarbeitung fuer die Medizin 1999, Informatik Aktuell, pages 216-220. Springer-Verlag, March 1999
9. Brennecke R, Burgel U, Burgel U, Simon R, Rippin G, Fritsch HP, Becker T, Nissen SE, *Measurement Of image quality differences at varying levels of data compression,* II Medical Clinic, Johannes-Gutenberg-University, Mainz, Germany, Eur Heart J 2000 Apr;21(8):687-696
10. Kerensky RA, Cusma JT, Kubilis P, Simon R, Bashore TM, Hirshfeld JW Jr, Holmes DR Jr, Pepine CJ, Nissen SE, *The effects of lossy data compression on recognition of diagnostic features in digital coronary angiography,* Cleveland Clinic Foundation, Cleveland, Ohio, Eur Heart J 2000 Apr;21(8):668-678
11. Tuinenburg JC, Koning G, Hekking E, Zwinderman AH, Becker T, Simon R, Reiber JH, *The effects of varying JPEG data compression levels on the quantitative assessment of the degree of stenosis in*

digital coronary angiography, Division of Image Processing (LKEB), Department of Radiology, LUMC, Leiden, The Netherlands, Eur Heart J 2000 Apr;21(8):679-686

12. Mitani M, Mitani T, Natori H, *Digital filing system for medical image data,* Department of Diagnostic Ultrasound and Medical Electronics, Sapporo Medical University School of Medicine, Rinsho Byori 1997 May;45(5):447-55

13. Koning G, van Meurs BA, Haas H, Reiber JH, *Effect of data compression on quantitative coronary measurements,* Department of Diagnostic Radiology and Nuclear Medicine, University Hospital, Leiden, The Netherlands, Cathet Cardiovasc Diagn 1995 Feb;34(2):175-85

14. Good WF, Maitz GS, Gur D, *Joint photographic experts group (JPEG) compatible data compression of mammograms,* Department of Radiology, University of Pittsburgh, PA 15261-0001, J Digit Imaging 1994 Aug;7(3):123-32

15. Sneiderman C, Schosser R, Pearson TG, *A comparison of JPEG and FIF compression of color medical images for dermatology,* National Library of Medicine, Bethesda, MD 20894, Comput Med Imaging Graph 1994 Sep-Oct;18(5):339-42

16. Kido S, Ikezoe J, Kondoh H, Takeuchi N, Johkoh T, Kohno N, Tomiyama N, Naito H, Arisawa J, Nakamura H, *Detection of subtle interstitial abnormalities of the lungs on digitized chest radiographs: acceptable data compression ratios,* Department of Radiology, Osaka University Medical School, Japan, AJR Am J Roentgenol 1996 Jul;167(1):111-5

Standardization of Image Collaboration Command Protocol for Telemedicine

[1]SHIGEJI HARA AND [2]TETSUO SAITO
[1]*College of Medical Technology, Fujita Health University Aichi, JAPAN* [2]*The Medical Information System Development Center, Tokyo, JAPAN*

Key words: Telemedicine, image collaboration, standardization, collaboration sever, token.

Abstract: In telemedicine, the standardization of fundamental operation commands and fundamental communication protocol is necessary to smoothly carry out image collaboration between many image collaboration systems developed by different companies. This paper reports the development of the standardization of image collaboration command protocol. The system model using a collaboration server was adopted to guarantee that all images displayed on the screen under collaboration were identical. In telepathology system, one terminal possesses a token, which issues the collaboration command to the other, in order to retain the identity of the image displayed on both terminals. Fabricating a function that receives and interprets a token into a collaboration server creates an image collaboration system that can be used in both teleradiology and telepathology. The result of field experiments showed that the image collaboration command protocol, which was standardized here, was useful for the medical care supports.

1. INTRODUCTION

In telemedicine, the performance of hardware such as display units and transmission equipment, and the security of information under image collaboration are very important. In addition to these requirements, the standardization of fundamental operation commands and fundamental communication protocol is necessary to smoothly carry out image collaboration between many image collaboration systems connected in the network, systems which have been developed by different companies. Until now, some image collaboration systems have been developed and utilized in the medical field[1] however, there is no example of image collaboration between different company's systems. MEDIS-DC planned to advance the standardization of image collaboration command protocol in order to collaborate between systems developed by different companies in the future. After receiving requests from MEDIS-DC, the IS&C committee asked members of 68 companies to cooperation in the standardization work. Then, the project WG9 was started to discuss standards.

Here, as a chief examiner who has managed WG9, the outline of the standard is reported in order to propagate the image collaboration command protocol described here. Also, results of field experiments [2,3] of the systems, which conform to this standard, are described briefly.

2. MODELING OF IMAGE COLLABORATION SYSTEM

In the image collaboration, it is necessary to guarantee that all images displayed on the screen under collaboration are identical. For this reason, exclusive control, which does not execute a new command before another one is finished, is necessary for image collaboration system. The model for this purpose is shown in Figure 1. Collaboration server receives a command from a terminal 1, and distributes the command to each terminal and also totally controls the image collaboration system. The function of the collaboration server can be realized by fabricating a software module in a terminal. This system using a collaboration server is a system for being excellent in respect of expansibility to 1 vs. N in the number of terminals.

In the meantime, image collaboration in telepathlogy (pathological diagnosis) is carried out between only two terminals. In such a system, one terminal possesses the authority (token), which issues the collaboration command to the other, in order to retain the identity of the image displayed on both terminals. This system using a token is economically constructed because of simple software technology.

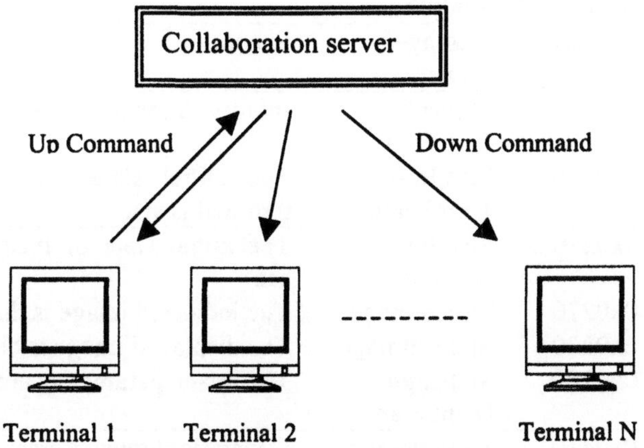

Figure 1. Collaboration server system

Fabricating a function that receives and interprets a token into a collaboration server creates an image collaboration system that can be used in both teleradiology and telepathology.

3. SCOPE

(1) Object image: Black and white or colour digitised medical images related to radiology, pathology, endoscopes and so on.
(2) Network form :One-to-one. However, the expansibility to 1 vs. N is considered.
(3) Range of the standard: Medical image and auxiliary information expressed in the image. Not including user interfaces.
(4) Object layer: On fifth or higher layer of the OSI 7-layer model.
(5) Security function: Using the security system utilized in ordinary telecommunication system.

Table 1 Collaboration Messages

Classification	Message ID	Message Name	Function
Collaboration, Condition	0x0020	SIquery-TermInfo	Inquiry of the terminal specification
	0x0030	SIsetInit-TermParam	The initial stage parameter of the terminal is set.
Image Display	0x0210	SIsetInit-DispParam	The initial state of image display is set.
	0x0220	SIdispImage	The indicated image is displayed.
	0x0230	SIcloseImage	The displayed image is closed.
	0x0240	SIchange-DispParam	The image parameters are changed.
Collaboration Annotation	0x0410	SItelePointer Msg	Generation/transfer/erasure of the pointer
	0x0420	SItele-OverlayMsg	Generation/transfer/addition/erasure of the rectangle binary image
	0x0430	SItele-DrawingMsg	Generation/transfer/addition/erasure of the drawing line
	0x0440	SItele-StringMsg	Generation/transfer/change/addition of the character string
	0x0450	SItele-EraseAllMsg	All annotations for collaborating except for the pointer is eliminated
	0x0460	SItelePixel-ValueMsg	Generation/transfer/erasure of grey level of a pixel
	0x0470	SIteleScale-Msg	Generation/transfer/erasure of the distance display between two points
Image Transmission	0x0780	SIstartImage-Trans	Start of the image transmission is notified.
	0x0790	SIsendImage-Data	Image datum is transmitted.
	0x07A0	SIendImage-Trans	The end of the image transmission is notified.
Telepathology	0x0B10	SIqueryMicro-Info	Inquiry of the microscope specification
	0x0B80	SIsendSlide-Image	Transmission of the glass sample whole image
	0x0B90	SisendMicro-Image	The transmission of the microscope image
Expanded Function	0x0930	SIreqToken	An acquisition demand of the token
	0x0940	SIgiveToken	The transfer of the token

4. IMAGE COLLABORATION COMMAND PROTOCOL

The main commands (=messages) defined by the image collaboration command protocol are shown in Table 1.

The main functions on the message of 0x0240 in Image Display shown in Table 1 are detailed as follows. These messages are considered to be used in teleradiology.

(1) The image display function: An indicated image file is read out and the identical image is displayed on each screen of all terminals.
(2) The grey scale change function: The function, which changes the grey scale of the displayed image.
(3) Scaling function: The function, which changes the scale of the displayed image.
(4) Rotating function: The function, which rotates the image in turns of 90 degrees.
(5) Image inversion function: The function, which inverts the image for right and left, top and bottom.
(6) Multi-frame display function: The function, which displays multiple images photographed in X-ray computed tomography inspection, etc.

The following messages are regulated as a peculiar message for telepathology.

(1) Query about the microscope specification (camera number, pixel number, extension magnification information, etc.)
(2) Transmission demand of the entire glass slide image
(3) Transmission demand of the microscopic image
(4) Token request: The terminal without a token requires the acquisition of a token.
(5) Token transfer: Transfers a token to the terminal without a token.

The collaboration messages are transmitted by the following procedures (Figure 2).

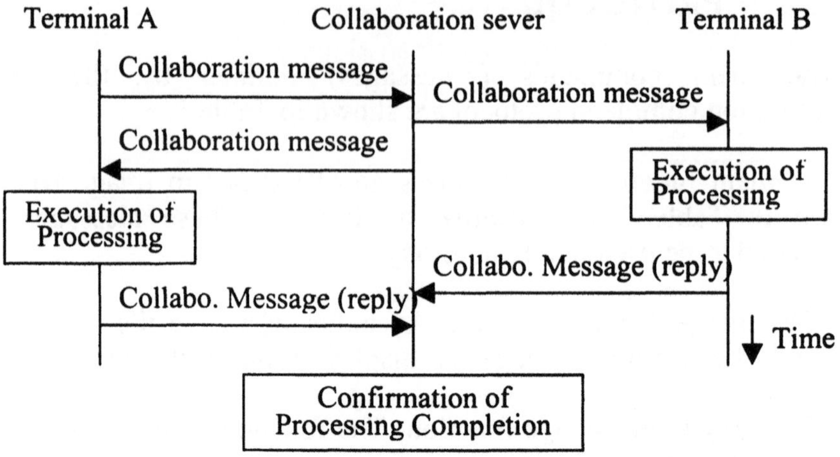

Figure 2. Basic procedure for collaboration message transmission

Step1: One terminal (terminal A) transmits a collaboration message to the collaboration server.

Step2: The collaboration server transmits the received collaboration messages to each of the terminals (A and B).

Step3: Each terminal that has received a collaboration message performs processing according to the contents of the message and transmits the processing results to the collaboration server as a reply message.

Step4: The collaboration server that has received the reply message confirms that processing on each terminal is completed.

The formats of image files used for both teleradiology and telepathology are basically in accordance with the DICOM (Digital Imaging and Communications in Medicine) standards. However, parts of the Supplements that are still under deliberation are cited. If the DICOM standards are changed in those deliberations, this format may also be reviewed accordingly.

5. FIELD EXPERIMENTS AND APPRECIATION

MEDIS-DC planned to evaluate the image collaboration system, which was built in to the image collaboration command protocol described above. Multiple regions which corresponded to remote islands and remote places in Japan were chosen, and clinics there and the core hospitals which supports them, were connected to the network by image collaboration systems, which were developed by multiple companies. Systems for teleradiogy were tested at the 5 fields of Kyoto, Okinawa, Iwate, Yamagata and Kagoshima. It was certified that pointer, marker and other annotations were synchronous on a screen of both request side and support side in spite of using image collaboration systems of different benders, and the systems were effective for medical care support.

Also systems for telepathology were tested between 3 fields in Okinawa, Kyoto, and Yamagata. Image transmission was mutually possible between systems of different benders, and the diagnosis support by a pathologist was possible using transmitted images. The result of these experiments showed that the image collaboration command protocol, which was standardized here, was useful for the medical care supports.

6. CONCLUSIONS

The standardization of image collaboration command protocol was discussed in order to collaborate between systems developed by different companies. The system model using a collaboration server was adopted to guarantee that all images displayed on the screen under collaboration were identical. The result of field experiments showed that the image collaboration command protocol, which was standardized here, was useful for the medical care supports.

Furthermore, the English version of these standards will be released with the command protocol installation standards in the near future[4].

ACKNOWLEDGEMENTS

The authors would like to thank all members of WG9 for their cooperation in the standardization work and would also like to thank the people concerned with the field experiments for their efforts.

REFERENCES

1. Takanobu Osaki, Hideyuki Ban, Hitoshi Matsuo, Akifumi Okabe, Kotaro Nakajima, Masahiro Yamaguchi, Nagaaki Ohyama, 1998, Remote diagnostic image support system with both the real-time system and the mail system, *MEDICAL IMAGING TECHNOLOGY* Vol.16 No.6 November; pp-615-621 et.

2. The Medical Information System Development Centre, 1999, *The remote place telemedicine system development reports*. (In Japanese)

3. The Medical Information System Development Centre and Information Technology Promotion Agency Japan, 2000, *Results of entrusted work from IPA for Medical Information System*, (in Japanese)

4. The Medical Information System Development Centre, 2000, *Standards for Image Collaboration Command Protocols* (under translation)

Index

List of Contributors

D. P. Auer	Arbeitsgruppe NMR, Institut für klinische Psychiatrie, Max-Planck-Institut München, F-80804 München, Germany.
L. M. Auer	Department of Medical Telematics and Robotics, Institute of Applied Sciences in Medicine (ISM), A-5020 Salzburg, Austria.
W. W. Baus	Institut für Radioonkologie/Strahlentherapie, Klinikum Darmstadt, Darmstadt, Germany.
V. Biefield	Institute for Microsensors, -Actuators and –System (IMSAS), University of Bremen, Germany.
J. Binder	Institute for Microsensors, -Actuators and –System (IMSAS), University of Bremen, Germany.
B. Blobel	Institute for Biometrics and Medical Informatics, Otto-von-Guericke University Magdeburg, D-39120 Magdeburg, Germany.
B. Clasbrummel	Berufsgenossenschaftliche Kliniken "Bergmannsheil", Bochum, Germany.
J. Ehrhardt	Institute for Medical Informatics, Medical University of Lübeck, D-23538 Lübeck, Germany.
K. Engel	Institute for Biometrics and Medical Informatics, Otto-von-Guericke University Magdeburg, D-39120 Magdeburg, Germany.
R. Feinen	Laufenberg GmbH & Co KG Bochum, Germany.
T. Fujii	Communications Research Laboratory, MPT, Koganei-city, Tokyo, Japan.
H. Handels	Institute for Medical Informatics, Medical University of Lübeck, D-23538 Lübeck, Germany.
S. Hara	College of Medical Technology, Fujita Health University, Aichi, Japan.
F. Holbe	Institute for Medical Psychology, Medical University of Lübeck, University Clinic Lübeck, D-23538 Lübeck, Germany.
H.-P. Howaldt	Department of Maxillo-Facial Surgery, Justus-Liebig-Univeristy, Medical School, D-35385 Giessen, Germany.
H. K. Huang	Department of Radiology, Childrens Hospital Los Angeles, University of Southern California, 4650 Sunset Boulevard, Los Angeles, California 90027, USA.
P. Hufnagl	Department of Pathology, Charité, Medical Faculty of the Humboldt University, Berlin, Germany.
B. Jettkant	Berufsgenossenschaftliche Kliniken "Bergmannsheil", Bochum, Germany.

H. Jiang	Communications Research Laboratory, MPT, Koganei-city, Tokyo, Japan.
T. H. Jünger	Department of Maxillo-Facial Surgery, Justus-Liebig-Univeristy, Medical School, D-35385 Giessen, Germany.
D. Keite	SerCon GmbH, Mainz, Germany.
T. Kobayashi	KGT Inc., Shinjuku-ku, Tokyo, Japan.
B. Kober	Institut für Radioonkologie/Strahlentherapie, Klinikum Darmstadt, Darmstadt, Germany.
D. Krechel	Knowledge-Based Systems Group, University of Kaiserslauten, D-67663 Kaiserslauten, Germany.
A. Kriete	Imagine Processing Laboratory, Justus-Liebig-Univeristy, Medical School, D-35385 Giessen, Germany.
F. Losemann	Institute für Telematik e. V, D-54292 Trier, Germany.
S. Loncaric	Faculty of Electrical Engineering and Computing, University od Zagreb, 10000 Zagreb, Croatia.
A. Margan	Practise for Internal Medicine, Cres, Croatia.
Ch. Meinel	Institute für Telematik e. V, D-54292 Trier, Germany.
T. Mitou	Department of Neurosurgery, Toho University School of Medicine, Tokyo, Japan.
A. Morgan	Department of Cardio-thoracic Surgery, Bristol Royal Infirmary, University of Bristol, Bristol, UK.
C. Otto	Department of Medical Informatics, Medical Office of the Bundeswehr, D-53225 Bonn, Germany.
E. Pek	Faculty of Electrical Engineering and Computing, University od Zagreb, 10000 Zagreb, Croatia.
P. Pharow	Institute for Biometrics and Medical Informatics, Otto-von-Guericke University Magdeburg, D-39120 Magdeburg, Germany.
S. Pieper	GMD, National Research Center for Information Technology, HCI Research Department, D-53754 Sankt Augustin, Germany.
S. J. Pöppl	Institute for Medical Informatics, Medical University of Lübeck, D-23538 Lübeck, Germany.
T. Posselt	Department of Maxillo-Facial Surgery, Justus-Liebig-Univeristy, Medical School, D-35385 Giessen, Germany.
F. Prämaßing	EPOS GmbH & Co. KG, an Infineon Technologies Company, Duisburg, Germany.
M. R. Rees	Department of Radiology-Division of Cardiac, Anaesthetic & Radiologic Sciences, University of Bristol, Bristol, UK.
B. Rehberg	SerCon GmbH, Mainz, Germany.
M. Reicherts	Department of Maxillo-Facial Surgery, Justus-Liebig-Univeristy, Medical School, D-35385 Giessen, Germany.
P. Richardson	Department of Clinical Radiology, Bristol Royal Infirmary, University of Bristol, Bristol, UK.
T. Saito	The Medical Information System Development Center, Tokyo, Japan.
S. Satou	KGT Inc., Shinjuku-ku, Tokyo, Japan.

M. Schmidt	EPOS GmbH & Co. KG, an Infineon Technologies Company, Duisburg, Germany.
F. Schmielau	Institute for Medical Psychology, Medical University of Lübeck, University Clinic Lübeck, D-23538 Lübeck, Germany.
H. Schmitz	Laufenberg GmbH & Co KG Bochum, Germany.
T. Schößler	Institute for Medical Informatics, Medical University of Lübeck, D-23538 Lübeck, Germany.
S. Seiwerth	Institute of Pathology, Faculty of Medicine, University of Zagreb, Croatia.
N. Shastry	Department of Radiology-Division of Cardiac, Anaesthetic & Radiologic Sciences, University of Bristol, Bristol, UK.
I. Shibata	Department of Neurosurgery, Toho University School of Medicine, Tokyo, Japan.
M. Silovic	Faculty of Electrical Engineering and Computing, University od Zagreb, 10000 Zagreb, Croatia.
W. Stefanon	Department of Medical Telematics and Robotics, Institute of Applied Sciences in Medicine (ISM), A-5020 Salzburg, Austria.
N. Sugou	Department of Neurosurgery, Toho University School of Medicine, Tokyo, Japan.
C. Tennstedt	Department of Pathology, Charité, Medical Faculty of the Humboldt University, Berlin, Germany.
L. Vorwerk	Institute für Telematik e. V, D-54292 Trier, Germany.
K. S. Wehrstedt	Department of Pathology, Charité, Medical Faculty of the Humboldt University, Berlin, Germany.
M. Weidenbach	Department of Pediatric Cardiology, University Bonn, D-53113 Bonn, Germany.
C. Wick	GMD, National Research Center for Information Technology, HCI Research Department, D-53754 Sankt Augustin, Germany.
E. K. Wong	Department of Ophthalmology, University of California Irvine, Medical Plaza, One Medical Plaza Drive, Irvine CA 92717, USA.

GPSR Compliance

The European Union's (EU) General Product Safety Regulation (GPSR) is a set of rules that requires consumer products to be safe and our obligations to ensure this.

If you have any concerns about our products, you can contact us on ProductSafety@springernature.com

In case Publisher is established outside the EU, the EU authorized representative is:

Springer Nature Customer Service Center GmbH
Europaplatz 3
69115 Heidelberg, Germany

Batch number: 10705495

Printed by Printforce, the Netherlands